AF847588

Smart Innovation, Systems and Technologies

Volume 168

Series Editors

Robert J. Howlett, Bournemouth University and KES International, Shoreham-by-sea, UK
Lakhmi C. Jain, Faculty of Engineering and Information Technology, Centre for Artificial Intelligence, University of Technology Sydney, Sydney, NSW, Australia

The Smart Innovation, Systems and Technologies book series encompasses the topics of knowledge, intelligence, innovation and sustainability. The aim of the series is to make available a platform for the publication of books on all aspects of single and multi-disciplinary research on these themes in order to make the latest results available in a readily-accessible form. Volumes on interdisciplinary research combining two or more of these areas is particularly sought.

The series covers systems and paradigms that employ knowledge and intelligence in a broad sense. Its scope is systems having embedded knowledge and intelligence, which may be applied to the solution of world problems in industry, the environment and the community. It also focusses on the knowledge-transfer methodologies and innovation strategies employed to make this happen effectively. The combination of intelligent systems tools and a broad range of applications introduces a need for a synergy of disciplines from science, technology, business and the humanities. The series will include conference proceedings, edited collections, monographs, handbooks, reference books, and other relevant types of book in areas of science and technology where smart systems and technologies can offer innovative solutions.

High quality content is an essential feature for all book proposals accepted for the series. It is expected that editors of all accepted volumes will ensure that contributions are subjected to an appropriate level of reviewing process and adhere to KES quality principles.

** **Indexing: The books of this series are submitted to ISI Proceedings, EI-Compendex, SCOPUS, Google Scholar and Springerlink** **

More information about this series at http://www.springer.com/series/8767

Isabella M. Lami
Editor

Abandoned Buildings in Contemporary Cities: Smart Conditions for Actions

Springer

Editor
Isabella M. Lami
Department of Regional and Urban Studies
and Planning (DIST)
Politecnico di Torino
Turin, Italy

ISSN 2190-3018 ISSN 2190-3026 (electronic)
Smart Innovation, Systems and Technologies
ISBN 978-3-030-35549-4 ISBN 978-3-030-35550-0 (eBook)
https://doi.org/10.1007/978-3-030-35550-0

© Springer Nature Switzerland AG 2020
This work is subject to copyright. All rights are reserved by the Publisher, whether the whole or part of the material is concerned, specifically the rights of translation, reprinting, reuse of illustrations, recitation, broadcasting, reproduction on microfilms or in any other physical way, and transmission or information storage and retrieval, electronic adaptation, computer software, or by similar or dissimilar methodology now known or hereafter developed.
The use of general descriptive names, registered names, trademarks, service marks, etc. in this publication does not imply, even in the absence of a specific statement, that such names are exempt from the relevant protective laws and regulations and therefore free for general use.
The publisher, the authors and the editors are safe to assume that the advice and information in this book are believed to be true and accurate at the date of publication. Neither the publisher nor the authors or the editors give a warranty, expressed or implied, with respect to the material contained herein or for any errors or omissions that may have been made. The publisher remains neutral with regard to jurisdictional claims in published maps and institutional affiliations.

This Springer imprint is published by the registered company Springer Nature Switzerland AG
The registered company address is: Gewerbestrasse 11, 6330 Cham, Switzerland

Preface

The issue of abandoned real estate asset has been the subject of wide-ranging debate over the last 20 years. Tackling this subject in order to trigger real operations of reuse implies the reconsideration of some beliefs. In particular, just to cite one example, the idea is that artificial capital is valid in the measure of its fruitfulness, while its usefulness depends on the political–administrative and sociocultural context that can favour processes of adaptive reuse.

The prospect of an innovative development of these properties, of these dormant capitals widespread in European cities, is justified precisely by the different considerations of their capital value. The types of value that can be created are called into question. I am referring not only to financial, but also to social and cultural ones, understood as values in themselves and not as a leverage.

The volume illustrates the results of a university research project on this theme, promoted and coordinated by me, entitled: "Shapes, Rules and Values in the contemporary city". The two-year research project was co-financed by the Interuniversity Department of Regional and Urban Studies and Planning (DIST) of the Politecnico and Università di Torino and has also involved the Polytechnic of Milan and IUAV of Venice. As the title of the research indicates, the study addressed the issue with an interdisciplinary approach that considered the role of typological and architectural characteristics, the regulatory framework and the aspects of value creation.

It is indeed the integration of different perspectives on the unused asset that has been of central concern to me in editing this book, fuelled by a belief that only addressing the issue in a non-ordinary way could trigger the reuse of urban abandoned spaces with low financial capital investment. Thus, during the discussion of theoretical aspects and evaluation methods, as a research group we have sought to illustrate them by selected case studies, but also by drawing attention to the practical issues of implementation.

I believe that the book can be of value to various reader groups, each with their own objectives:

- Public administrators, who are currently regulating urban transformations mainly through agreement instead of imposition, sincerely willing to determine the quality of the exchanges and, ultimately, legitimise the agreement between public and private stakeholders;
- Entrepreneurs, architects and planners, who want to activate recovery and enhancement operations, economically feasible and culturally stimulating;
- Academics with various backgrounds that could contribute to the debate.

I wish you all as much enjoyment and satisfaction in reading this book as we, as a research group, had in preparing it.

Turin, Italy Isabella M. Lami

Acknowledgement I would like to acknowledge financial support from the Department of Regional and Urban Studies and Planning (DIST) of the Politecnico di Torino, which co-funded this research.

Contents

Part III Discussion and Conclusions

Chapter 1
Introduction

Isabella M. Lami

1.1 Aim of the Book

The purpose of this book is to study the conditions for activating reuse operations of urban abandoned spaces, with low intensity of financial capital and high intensity of cultural and human capital, through a new culture of design and regulation.

The starting point is the relevant phenomenon in the contemporary city, represented by the existence of underused or vacant buildings and spaces. These buildings embody a dormant asset that furthermore creates problems and degradation in the urban fabric. The reality of several European cities shows how traditional forms of stimulating urban renewals (with respect to the financing of operations, how to design and build, and urban planning legislation) no longer work. While it is clear that the values to which we are referring are multidimensional, the relationships among the various dimensions are still obscure.

This book analyses the phenomenon of unused and abandoned buildings based upon a series of elements in the Italian context, where a nation divided into two parts can be observed: the (big) cities and the rest of the country which is gradually becoming depopulated (Micelli and Pellegrini 2018). This book focuses on the former, where, even though the phenomenon of vacant buildings is not so dramatic, it is, however, of considerable magnitude. We decided to study cities because they are the only context in which it is realistically possible to envisage the triggering of reuse phenomena, for a number of reasons which we illustrate in this book.

It must be noted that in Italy there is no single database containing up-to-date information on the consistency of these assets: such an absence reveals that the attention given to this potential capital is still marginal, despite the broad debates on the matter. This gap also emerges in the book, in which the data provided in the

I. M. Lami (✉)
Department of Regional and Urban Studies and Planning (DIST), Politecnico di Torino,
Viale Mattioli 39, 10122 Turin, Italy
e-mail: isabella.lami@polito.it

© Springer Nature Switzerland AG 2020

I. M. Lami (ed.), *Abandoned Buildings in Contemporary Cities: Smart Conditions for Actions*, Smart Innovation, Systems and Technologies 168,
https://doi.org/10.1007/978-3-030-35550-0_1

following chapters may not be homogeneous, according to the different databases and references used, as explained below.

Firstly, the 2011 national census (reported in ISTAT 2014) identifies a total of more than 14 million buildings in Italy, of which approximately 12 million buildings for residential use and about 2 million are used for other public and private purposes; 5.2% of the total assets are in a state of disrepair, collapsing or under construction, corresponding to 754,821 buildings.

These approximately 12 million buildings are equivalent to a total number of more than 31 million of dwellings (i.e. "spaces- or sets of spaces- permanently intended for residential use; surrounded by walls and covered by a roof; independent and inserted in a building—or which constitutes a building itself"), 7 million of which are "empty or occupied by non-residents".

Moroni et al., the authors of the second chapter of this book, assume a total of more than 2,500,000 vacant Italian dwellings. The data is obtained from the figures mentioned above and on the basis of the studies of Gentili and Hoekstra (2018), according to which only 40% of dwellings classed as "empty or occupied by non-residents" are truly empty, and are integrated with data provided by The Guardian (2014).

According to the studies reported by Campagnoli and Tognetti (2016), the total number of unused assets is over six million, including both buildings and dwellings. Table 1.1 provides details of the unused buildings with respect to their intended use, presented and discussed in the chapters by Micelli and Mangialardo (third) and Bonini Baraldi and Salone (sixth).

It is not easy to compare between the two sources, as the definitions of vacant/unused/occupied are slightly different, but it seems plausible to state that there are between approximately 5 and 7 million unused dwellings, equal to about 16–22% of the dwellings present in the country. This is an impressive portion which is destined to increase, given the demographic trend and the economic crisis, if action is not taken through targeted urban policies.

Table 1.1 Overview of abandoned buildings distinguished by uses

Industrial sheds	700,000
Empty offices and shops	650,000
Abandoned railway stations	1900
Assets available from the State	16,499
Assets seized from mafia organizations	13,118
Buildings of consortia, public bodies and companies	1244
Abandoned buildings of historical/cultural value	20,000
Non-residential Buildings (total)	1,402,761
Unused dwellings (total)	5,000,000

Source Campagnoli and Tognetti (2016)

To address the issue in a non-ordinary way (i.e. with a simple logic of real estate development), we must refer to four key concepts of the current debate in the field of urban studies and urban economy (necessarily measured by the ongoing conditions of austerity that characterise the current urban policy), attempting to integrate them within the framework of an innovative concept of urban development in the European context:

- Adaptive reuse: a radical approach to the reuse of architectural and urban spaces, namely adapting the new function to the host space, and not vice versa, keeping the original features unaltered (Németh and Langhorst 2014; Robiglio 2016);
- Circular economy: it can be defined as "restorative and regenerative by design. The transition to renewable energy sources, the circular model builds economics, natural and social capital" (Ellen MacArthur Foundation 2017);
- Sharing economy: it does not have a single definition, but can be described as "an economic system based on sharing underused assets or services, for free or for a fee, directly from individuals" (Botsman and Rogers 2010);
- Finally, the notion of cultural capital can offer many different definitions, starting with the contribution of Bourdieu (1986), which focuses on the individual endowment of cultural capital, and going so far as the vision of Throsby (1999) which distinguishes between tangible and intangible cultural capital. Here, we need to adopt new interpretative strategies, in which the concept of cultural capital is broadened and merged with that of human capital, seen as the set of skills, knowledge, expertise, relationships and abilities possessed by the individual and the community in a broad sense; developed through a personal and collective journey.

1.2 Outline of the Book

The starting point of the book was a research co-financed by the Department Dist of the Politecnico di Torino, Italy, entitled "Shape, Rules and Values". Due to the fact that the research question ("How are operations of reuse of urban abandoned spaces activated with new approaches?") requires an interdisciplinary approach, the research group was made up of experts in urban evaluation, design, planning, management, geography and economics from four Academies in Italy (Politecnico di Torino, University of Turin, Polytechnic of Milan, and Iuav).

The structure of the book reflects the richness and variety of the approaches to the topic, and the intention to carry out an integrated study that observes the phenomenon in its economic, architectural and urban issues.

Starting from the theoretical aspects of the topic, the book offers an analysis of six Italian case studies, diversified in terms of geographical distribution and in the governance model, but united by a common thread represented by the presence of an important role in the cultural destination for the reuse proposal. The six case studies investigated are: Toolbox, Turin; Factory Grisù, Ferrara; Caos, Terni; Officine Zero,

Rome; Ex Fadda, San Vito dei Normanni (Brindisi) and Farm Cultural Park, Favara (Agrigento).

The book is split into three main sections:

I. The first part is theoretical and provides an overview of the relevant concepts in the field of the reuse of vacant buildings;
II. The second part is devoted to case studies, with a description of the case selection and the analysis methods, a detailed report on the cases, a series of comparative reflections on the six interventions;
III. The final chapter connects the two parts, reworking the theoretical aspects starting with empirical studies and proposing generalised reflections.

(I) In the first part of the theoretical contribution, the main issues of the vacant buildings are analysed. Chapter 2 (Moroni et al.) gives an answer to three fundamental questions: (i) *definitions* (i.e. precisely what phenomena are we dealing with?), (ii) *problems* (i.e. when and why do certain phenomena represent a publicly significant issue?), and (iii) *strategies* (i.e. what needs to be done to limit the negative aspects of certain phenomena and to set positive changes in motion?). Chapter 3 (Mangialardo and Micelli) argues that public real estate assets have changed their role in the urban agenda, ceasing to be a simple reserve of useful value to compensate the financial difficulties of the Administrations, being transformed into a resource capable of hosting social and economic innovation. The fourth Chapter (Baima and Robiglio) proposes, through the concept of intensity, a different point of view for approaching new questions posed by the contemporary urban context, concerning the reactivation and recoding of uses and practices within consolidated city spaces. Chapter 5 (Mecca and Lami) summarises some of the main issues of the evaluations of urban regenerations. Based upon the literature review of the last twenty years, which reveals little literature supporting the method of assessing value creation in relation to urban regeneration, the aim of the chapter is to provide an evaluation proposal to assess the outputs of a regeneration process based on culture.

(II) The section of case studies is opened by the Bonini Baraldi and Salone chapter (the sixth), which describes the research methodology, a hybrid method that combines different disciplinary approaches—evaluation methods, urban studies approaches, and management science—in a field analysis based on qualitative and quantitative data. Chapter 7 (Porta and Abastante) aims to illustrate the six adaptive reuse cases according to the following labels: local context, architectural characters, origin and stakeholders, pills of history and value proposition. Chapter 8 (Bonini Baraldi and Salone) summarises the major findings emerging from the six case studies, highlighting three aspects: (i) Genesis, governance and organisation; (ii) business model and economic sustainability and (iii) socio-spatial relations.

(III) The final chapter (Lami) provides a series of theoretical insights from empirical research observations of the cases analysed in the field. The reflections concern, in particular, the need, in a territory with "zero profitability", to develop enabling policies for adaptive and creative reuse of unused real estate assets.

References

Campagnoli, G., & Tognetti R. (2016). *L'Italia da riusare, La Nuova ecologia*. www.osservatorioriuso.it. Accessed on May 10, 2019.

Botsman, R., & Rogers, R. (2010). *What's mine is yours. The rise of collaborative consumption*. New York: Harper Collins Publishers.

Bourdieu, P. (1986). Forms of Capital. In John G. Richardson (Ed.), *Handbook of theory and research for the sociology of education*. New York: Greenwood.

Ellen MacArthur Foundation, McKinsey Centre for Business and Environment. (2017). *Cities in the circular economy: An initial exploration*. Report published by Ellen MacArthur Foundations https://www.ellenmacarthurfoundation.org/assets/downloads/publications/Cities-in-the-CE_An-Initial-Exploration.pdf. Accessed on April 26, 2019.

Gentili M., & Hoekstra J. (2018). Houses without people and people without houses: A cultural and institutional exploration of an Italian paradox. *Housing Studies*, 1–23.

ISTAT. (2014). *Edifici e abitazioni*. Available at https://www.istat.it/it/files//2014/08/Nota-edifici-e-abitazioni_rev.pdf. Accessed on June 3, 2019.

Micelli, M., & Pellegrini, P. (2018). Wasting heritage. The slow abandonment of the Italian Historic Centers. *Journal of Cultural Heritage, 31*, 180–188.

Neate, R. (2014). Scandal of Europe's 11 m empty homes. *The Guardian*. https://www.theguardian.com/society/2014/feb/23/europe-11m-empty-properties-enough-house-homeless-continent-twice. Accessed on June 5, 2019.

Németh, J., & Langhorst, J. (2014). Rethinking urban transformation: Temporary uses for vacant land. *Cities, 40*, 143–150.

Robiglio, M. (2016). *The adaptive reuse toolkit. How cities can turn their industrial legacy into infrastructure for innovation and growth*. Urban and regional Policy Paper 38.

Throsby, D. (1999). Cultural capital. *Journal of Cultural Economics, 23*, 3–12.

Part I
Perspectives

Chapter 2
Vacant Buildings. Distinguishing Heterogeneous Cases: Public Items Versus Private Items; Empty Properties Versus Abandoned Properties

Stefano Moroni, Anita De Franco and Beatrice Maria Bellè

Abstract The issue of unused buildings figures high on the agenda in many countries. However, a clear-cut distinction is not always drawn between completely different cases: for instance, public and private buildings, or private buildings that are merely empty as opposed to private buildings that are abandoned. This chapter discusses these aspects critically.

2.1 Preface

The issue of unused buildings figures high on the agenda in many countries, including Italy, sparking heated public and academic debate. However, a clear-cut distinction is not always drawn between cases which we consider to be completely different: for instance, *public* and *private* buildings, or private buildings that are merely *empty* as opposed to private buildings that are *abandoned*. In this chapter we shall discuss these aspects critically. In so doing, we shall pay particular attention to the Italian situation, focusing particularly on urban settings.

2.2 Three Questions: Definitions, Problems, Strategies

We shall consider three main questions: (i) *definitions* (i.e. precisely what phenomena are we dealing with?), (ii) *problems* (i.e. when and why do certain phenomena

S. Moroni (✉) · A. De Franco · B. M. Bellè
Department of Architecture and Urban Studies (DASTU), Polytechnic University of Milan, via Bonardi 3, Milan, Italy
e-mail: stefano.moroni@polimi.it

A. De Franco
e-mail: anita.defranco@polimi.it

B. M. Bellè
e-mail: beatricemaria.belle@polimi.it

© Springer Nature Switzerland AG 2020
I. M. Lami (ed.), *Abandoned Buildings in Contemporary Cities: Smart Conditions for Actions*, Smart Innovation, Systems and Technologies 168,
https://doi.org/10.1007/978-3-030-35550-0_2

represent a publicly significant issue?), and (iii) *strategies* (i.e. what needs to be done to limit the negative aspects of certain phenomena and set positive changes in motion?).

2.2.1 First Question: Definitions

Different terms are used in different languages to denote the phenomena with which we shall be dealing here. Moreover, these terms are not always transparent or unambiguous. It is, therefore, crucial first to clarify certain basic definitions. It is particularly important to distinguish clearly between buildings that are merely *empty* and buildings that are *abandoned*.

A merely empty building or dwelling is simply a building or dwelling in good maintenance conditions but that is not used. In other words, no particular activity takes place within it.[1] Hence a holiday home used occasionally *is not* an empty building in this sense.

An abandoned building is a specific type of unused building: that is, an unused building which is not wholly functional or serviceable. In other words, it is a building lacking management and maintenance. In this case, windows and doors are not in perfect working order, components of the facade or roof pose a hazard, the electrical, plumbing or gas systems are not in a fully safe and functional state.[2] Abandoned buildings, as thus defined, are therefore buildings whose owners do not comply, either wholly or in part, with the responsibilities connected with the status of ownership (Hillier et al. 2003; Mallach 2006). Indeed, being an owner implies, ultimately, both *rights* and *duties* (Shoked 2014). Certain critics of private property tend to play down the former, while certain defenders of private property tend to underestimate the importance of the latter.

It is interesting that many alarming estimates of the phenomena discussed here do not distinguish clearly between the two above-mentioned cases (which are—for

[1]Considering the most recent ISTAT census (2014) and assuming (Gentili and Hoekstra 2018) that around 40% of the dwellings classed as "empty or occupied as other than primary residence" are truly empty, there are more than 2,500,000 empty dwellings in Italy according to our definition. This figure is comparable to the one given in a recent report by *The Guardian*: see https://www.theguardian.com/society/2014/feb/23/europe-11m-empty-properties-enough-house-homeless-continent-twice (accessed February 2019).

[2]It is not easy to estimate the number of abandoned buildings (thus defined) in Italy. According to ISTAT data (2014), the percentage of the total building stock which is unused because it is "collapsing, in a state of disrepair or under construction" is 5.2%, giving a total of around 750,000 buildings. The problem is that, on the one hand, it is not possible to separate the figure for buildings under construction from the total and, on the other, that the focus is on situations of severe neglect. Another source is the Italian Agenzia delle Entrate (2017b), which estimates around 474,000 properties defined as "*collabenti*" (i.e. reduced to ruins). In this case the figure does not include buildings under construction, yet still focuses exclusively on situations of severe neglect.

example in Italy and, particularly, in urban settings—on totally different scales).[3] Sometimes an even broader, vaguer category is used to accentuate the situation dramatically: that of buildings or dwellings "at the owner's disposal" (including not only truly empty and abandoned buildings, but also those that are used occasionally, for example holiday homes).[4] Here, we assume that the category of buildings or dwellings "at their owners' disposal" is, in contrast, negligible. Hence we do not take it into consideration.[5]

2.2.2 *Second Question: Problems*

In considering *whether* and *why* empty or abandoned buildings can create problems, we believe that it is essential to distinguish clearly between public buildings and private ones.[6] This distinction, strictly speaking, is based on the ownership of the assets. It separates assets owned by a public authority (for example a ministry, a region or a municipality)[7] from those owned by private citizens (a family, a company, etc.).[8]

Let us focus first on *public* buildings. Empty public buildings (i.e. unused public buildings) are a problem by definition because they are a public "resource", belonging to everyone, which is frozen and underused.[9] They thus represent a form of wastage of shared resources. This is an issue that has generally received scant attention in both theoretical (Tanzi and Prakash 2000) and practical terms (Kaganova and Nayyar-Stone 2000).

As Kaganova and Nayyar-Stone (2000, p. 307) write: "Although municipal governments generally attend—with greater or less experience—to the day-to-day operational needs of their real estate holdings, few such governments think of their holdings as a 'portfolio' whose composition might be modified to better serve public purposes. Nor does the typical municipality routinely review if the current use of

---

[3]See previous footnotes. Obviously, in this case as in others, any estimate depends to a large extent on how the phenomenon being estimated is defined (Morckel 2014).

[4]The Eurostat report (2015), for example, points out with some concern that almost one in six dwellings in the EU is unoccupied. But, according to this report, dwellings are classified as *unoccupied* "if they are reserved for seasonal or secondary use (such as holiday homes) or if they are vacant (dwellings which may be for sale, for rent, for demolition, or simply lying empty and unused)" (Eurostat 2015: 75).

[5]According to the Italian Agenzia delle Entrate (2017a), properties "at their owners' disposal but not used on a continuous basis" in 2014 accounted for 11.7% of properties owned by natural persons. Dwellings "at their owners' disposal" accounted for over 20% of the total in 3630 municipalities.

[6]These two main categories naturally include other sub-categories (Chiodelli and Moroni 2014). However, it is not our intention here to enter into greater details on this point.

[7]An overview and quantification of non-financial assets owned by public authorities in Italy can be found in MEF (2015).

[8]An overview and quantification of non-financial assets owned by households and non-financial companies in Italy can be found in Banca d'Italia (2019).

[9]In 2015, the value of unused public buildings in Italy was estimated at 12 billion euros (MEF 2015), for a total of 19 million square metres (Carapella 2018).

individual properties is appropriate from the perspective of opportunity cost, mode of management and finance, or match its long-term needs for property for its own use and as investment" (with specific reference to Italy, see Vermiglio 2011).

The point here is that the state is an organisation with precise, formally predetermined purposes: specifically, in the case in question, to use the assets and resources in its possession for the good of its citizens. It is, therefore, a "single owner" with unitary aims. In this case, it makes sense to speak of efficiency in how predetermined (public, shared) aims are pursued.[10] Clearly, if public buildings are not merely unused but are also abandoned, they raise further problems in terms of risks to people and things and in terms of various negative impacts.

Let us now consider *private* dwellings and buildings. Empty, unoccupied private spaces which nevertheless undergo ordinary, regular maintenance are not a problem *in themselves*. Firstly, they are privately-owned property; whether or not their owner uses them, profits from them or not, is publicly irrelevant.[11] Furthermore, a city without empty buildings could not function; it would be totally inflexible and unable to embrace either change or new residents and enterprises (see pp. 10–33 in Evans and Hartwich 2005).[12]

As Gentili and Hoekstra (2018) point out, the existence of large numbers of empty dwellings has often been considered a problem of "social justice" in the circumstance of housing shortage. However, the fact that there are empty private buildings or dwellings while there are people who do not have a home is of little public relevance. Naturally we *can* and *must* be concerned about people without homes[13] (and come up with *other* forms of public measures to help them). Yet there is no functional or ethical relationship between the two aforementioned phenomena. The city, as a socio-material system, is not in fact an organisation (i.e. a made order) in which all roles and positions can be integrated, compensated for and counterbalanced from an efficiency-based perspective. The city, in reality, is a complex dynamic order (Ikeda 2007; Gordon 2012; Holcombe 2013; Palmberg 2013); that is, a polycentric emergent social-spatial order whose (i) separate, independent members (with legitimately different goals) (ii) interact in mainly indirect, unpredictable ways, (iii) relying (for their life plans) on goods and resources which they legitimately acquire and use without violating applicable laws (Moroni 2015a). In other words, it is not possible

[10]Of course, efficiency is not *everything* in this case, but cannot be ignored.

[11]Unless, obviously, it is the existence of inadequate public rules that makes, for instance, renting difficult.

[12]Could we say that the existence of empty private dwellings and buildings generates systemic problems only above a certain threshold (i.e. percentage), as discussed by Glock and Häussermann (2004) and Huuhka (2016)? But how should such a threshold be calculated? In light of what economic theory or model? And what (empirical and/or evaluational) significance should be attributed to it? Moreover, would it be an identical threshold for each city (whether small, medium-sized or large) and for different national contexts?

[13]For an overview of the situation in Europe, see FEANTSA (2018).

to speak of efficiency when considering an *entire* city as a complex socio-economic system (Ikeda 2010, 2017).[14]

In the case of private buildings, the central (public) problem is not so much that there are unused, unoccupied buildings as that some of them are abandoned. Abandoned private buildings are a problem because (i) they can cause collapse and direct harm to people and things, (ii) they can have negative effects on the neighbourhood in which they are located (for instance, disinvestment in the neighbourhood and reduction in the value of neighbouring buildings), and (iii) they can attract and host illicit criminal activities. As Cohen (2001, p. 416) observes: "Although it is true that abandoned homes are symptomatic of other problems, they also contribute to neighborhood decline and frustrate revitalization efforts by becoming eyesores, fire hazards, and sites for drug-related activities, vagrancy and rodent infestation".

2.2.3 *Third Question: Strategies*

Obviously, the question of policies is entirely different in the two cases under consideration: that is, public and private buildings.

Let us begin by considering *public* assets. Three main strategies seem possible here.

Firstly, unused public buildings can be sold on the market. In a country like Italy, however, one cannot expect this to have a major impact on public debt, as is sometimes supposed, since the latter has reached extremely high levels and in any case such sales are one-time only (Cottarelli 2016).[15] The sale of unused buildings may nevertheless relieve public authorities of maintenance and management costs and guarantee a certain revenue. The sale of public buildings often raises doubts and provokes criticisms (because, for example, the public portfolio would be impoverished as a whole, because it would mean selling off the "family jewels", and so on: Gaeta and Savoldi 2013). Yet, given the enormous number of publicly-owned buildings in Italy[16] and the difficulty of keeping them *all* in use, these concerns appear to be justifiable only with regard to a part of the building stock in question (particularly buildings of clear cultural interest; that is, pertaining to the country's cultural heritage: Benedikter 2004). Naturally, the concrete possibility of selling certain buildings will depend on the actual demand on real estate market, which at the time of writing (2019) is not particularly strong.

---

[14]It should be noted that, paradoxically, it is precisely those who criticise the efficiency-based logic of the market in the name of an idea of social justice (applied by linking the various individual positions *linearly* and *directly*: in particular, home owners and homeless people) who end up by adopting the most banal efficiency-based perspective.

[15]On the management of public debt and the role of non-financial public assets in this regard, see Rawdanowicz et al. (2011) and Bova et al. (2013).

[16]In Italy, there are around one million publicly-owned buildings, totalling around 325 million square metres (MEF 2015). Over 70% of these assets are the property of local authorities.

Secondly, unused public buildings could be used as a source of income. This is often feasible for merely empty buildings in good condition; naturally, a solid endowment of entrepreneurialism (which is often lacking in Italy) must be possessed by the public authority concerned. For abandoned buildings the matter is more complex, because it would require major preliminary repurposing operations, which is no easy task given the current lack of funds at the disposal of Italy's public administrations.[17]

Thirdly, unused public buildings may be granted to civic associations and organisations for their use free of charge, on condition that they look after them. This is an option that is increasingly common in Italy as elsewhere, and it is generally viewed in especially positive terms. It would appear necessary, however, to strip this option of a certain recurrent rhetoric which maintains that it intrinsically coincides with social innovation and nearly always credits it with social effects that extend beyond the groups directly involved.[18] It would be better instead to consider it just one opportunity among others, where there are reciprocal benefits—for the public institution, as representative of all citizens, and for certain types of civic actors. It is also important to stress that this solution can only work if certain specific conditions apply (Micelli and Mangialardo 2016; Mangialardo and Micelli 2016, 2018), and in particular (i) if the urban area in question is characterised by strong social capital, (ii) if true "social entrepreneurs" are involved,[19] and (iii) if the situation allows the associations and organisations involved to obtain some kind of profit from some of the activities initiated.[20]

The crucial point to stress here is that, if the a priori ideological prejudices that often inform the current debate in Italy are avoided, the three above-mentioned measures can be part of the same general strategy, without necessarily being mutually exclusive. Indeed, there will be buildings and situations which are more suitable for one solution than another (according to their location and physical form, among other things). No single solution is better or worse than others *in itself*, while all of them may be useful in tackling the problem of unused public buildings.

We shall now consider *private* buildings and dwellings. In the case of private situations, the question of public measures primarily concern abandoned buildings and dwellings more than—as has already been stressed—empty buildings as such.[21]

---

[17]On methods and strategies to manage public real-estate portfolios appropriately, see e.g. Kaganova and Nayyar-Stone (2000), Vermiglio (2011), Trojanek (2015), Marona and van den Beemt-Tjeerdsma (2018), Constantin et al. (2018).

[18]See on this the critical discussion in Vitale (2009). On social innovation more generally, see recently, with specific reference to Italy, Caroli (2018).

[19]As happened for example in the cases of the *Ex Fadda* project in Brindisi and *Farm Cultural Park* in Favara (see the other chapters in this book).

[20]It should be noted here that, under Italian law, formally non-profit activities may make some profit provided that it does not become the main purpose and scope of the association in question.

[21]We therefore do not consider measures that target merely empty buildings. From our perspective it appears neither plausible nor desirable to envisage a special, higher tax rate for merely empty houses. Nor does it seem possible to regulate empty buildings in any specific way insofar as they are empty (see on this Hirokawa and Gonzalez 2010).

In this case there are two sides to the issue (we stress once again that here we are referring especially to urban settings).

Firstly, in cases where the state of abandonment constitutes a risk to other people's properties or to people themselves (for example due to the potential collapse of components or parts of a building), local authorities should be able to enjoin private owners to take action; that is, oblige them to restore the building in question to a safe condition. This is already possible to some extent in Italy and other countries. It should, however, form part of a more structured, systematic and stringent approach than is currently the case. In this regard, on the one hand, local building codes should be stricter and more specific in requiring adequate levels of building safety and maintenance. On the other hand, a more radical, active policy of code enforcement needs to be adopted (Kraut 1999).[22] It should be noted that in such cases, the action taken would not be *against* but rather *in favour of* private ownership as a right understood in its general sense. The safeguarding of private property in fact implies the safeguarding of a general situation in which *no* single property should cause direct or tangible damage to others (Moroni 2018).

Secondly, in order to encourage the repurposing of abandoned buildings in more general terms (that is, not simply making them safe again but also reintegrating them into urban life), several enabling policies could be implemented. For example: (i) certain current restrictions on the use, or change of use, and certain non-essential standards (that is, ones which do not regard fundamental questions of safety and security or well-defined nuisances) could be eliminated (Goldstein et al. 2001; Brophy and Vey 2002; Olivadese et al. 2017), (ii) certain complex bureaucratic procedures could be radically simplified (Adair et al. 2000, 2003); (iii) particular incentives could be introduced (for instance, increases in the surface area or volume of the building if it is renovated, and is renovated according to certain predetermined conditions, or subsidies for modernisation interventions: Glock and Häussermann 2004; Hollander 2009); (iv) taxation could be reduced on improvements made to buildings and conversely increased on land value (Vincent 2012; Farris 2015; Foldvary and Minola 2017; Moroni and Minola 2019).

2.3 Final Remarks

The question of unused buildings (in the various forms considered) is undoubtedly urgent. In order to tackle it in an effective and constructive manner, however, it is necessary to clarify a number of key points and avoid a priori prejudices. The aim of this chapter has been to present certain pivotal issues in the discourse which warrant renewed critical attention. Of course, within the limited space of a chapter, it has

[22]All of this suggests that it is necessary to rethink the crucial role of local building codes, after decades—particularly in Italy—during which they have been weighed down with rules regarding certain entirely secondary, negligible aspects (Moroni 2015b). To be noted in this regard is that building codes have generally received scant attention in the academic literature (Moroni 2012), including economic studies (as Van Doren 2005, points out).

not been possible to examine those individual issues in detail. The hope, however, is that our observations may at least signal the direction that seems most profitable and promising for research and debate.

References

Adair, A., Berry, J., McGreal, S., Deddis, B., & Hirst, S. (2000). The financing of urban regeneration. *Land Use Policy, 17*(2), 147–156.
Adair, A., Berry, J., & McGreal, S. (2003). Financing property's contribution to regeneration. *Urban Studies, 40*(5–6), 1065–1080.
Agenzia delle Entrate. (2017a). *Gli immobili in Italia.* Available at http://www.mef.gov.it/documenti-pubblicazioni/altri-documenti/index.html#cont6. Accessed December 2018.
Agenzia delle Entrate. (2017b). *Statistiche catastali 2016.* https://wwwt.agenziaentrate.gov.it/mt/osservatorio/Tabelle%20statistiche/Statistiche_Catastali_2016_12072017.pdf. Accessed December 2018.
Banca d'Italia. (2019). *La ricchezza delle famiglie e delle società non finanziarie italiane 2005–2017.* Available at https://www.bancaditalia.it. Accessed May 2019.
Benedikter, R. (2004). Privatisation of Italian cultural heritage. *International Journal of Heritage Studies, 10*(4), 369–389.
Bova, M. E., Dippelsman, M. R., Rideout, M. K. C., & Schaechter, M. A. (2013). *Another look at governments' balance sheets: The role of nonfinancial assets.* IMF working papers, no. 13/95. International Monetary Fund.
Brophy, P. C., & Vey, J. S. (2002). *Seizing city assets: Ten steps to urban land reform.* Washington: Brookings Institution Center on Urban and Metropolitan Policy.
Carapella, P. (2018). *Il patrimonio immobiliare della pubblica amministrazione.* Available at https://osservatoriocpi.unicatt.it/cpi-Patrimonio_immobiliare_pa_new.pdf. Accessed December 2018.
Caroli, M. G. (Ed.). (2018). *Evidenze sull'innovazione sociale e sostenibile in Italia.* Milano: Angeli.
Chiodelli, F., & Moroni, S. (2014). Typology of spaces and topology of toleration: City, pluralism, ownership. *Journal of Urban Affairs, 36*(2), 167–181.
Cohen, J. R. (2001). Abandoned housing: Exploring lessons from Baltimore. *Housing Policy Debate, 12*(3), 415–448.
Constantin, D. L., Mitrut, C., Grosu, R. M., Profiroiu, M., & Iosif, A. E. (2018). Municipal real properties and the challenges of new public management. *International Review of Administrative Sciences, 84*(1), 122–143.
Cottarelli, C. (2016). *Il Macigno.* Milano: Feltrinelli.
Eurostat. (2015). *People in the EU: Who are we and how do we live?* Available at https://ec.europa.eu/eurostat. Accessed April 2019.
Evans, A., & Hartwich, O. (2005). *Unaffordable housing.* London: Policy Exchange.
Farris, N. (2015). What to do when main street is legal again: Regional land value taxation as a new urbanist tool. *University of Pennsilvanya Law Review, 164,* 755–777.
FEANTSA. (2018). *Third overview of housing exclusion in Europe.* Available at https://www.feantsa.org. Accessed May 2019.
Foldvary, F. E., & Minola, L. A. (2017). The taxation of land value as the means towards optimal urban development and the extirpation of excessive economic inequality. *Land Use Policy, 69,* 331–337.
Gaeta, L., & Savoldi, P. (Eds.). (2013). *Orientamenti per la gestione del patrimonio immobiliare pubblico.* Società Italiana degli Urbanisti.
Gentili, M., & Hoekstra, J. (2018). Houses without people and people without houses: a cultural and institutional exploration of an Italian paradox. *Housing Studies,* 1–23. https://doi.org/10.1080/02673037.2018.144709.

Glock, B., & Häussermann, H. (2004). New trends in urban development and public policy in eastern Germany: Dealing with the vacant housing problem at the local level. *International Journal of Urban and Regional Research, 28*(4), 919–929.

Goldstein, J., Jensen, M., & Reiskin, E. (2001). *Urban vacant land redevelopment: Challenges and progress*. Working Paper. Cambridge (Mass.): Lincoln Institute of Land Policy.

Gordon, P. (2012). Spontaneous cities. In D. Andersson (Ed.), *The spatial market process* (pp. 181–209). Emerald: Bingley.

Holcombe, R. G. (2013). Planning and the invisible hand: Allies or adversaries? *Planning Theory, 12*(2), 199–210.

Hollander, J. B. (2009). *Polluted & dangerous: America's worst abandoned properties and what can be done about them*. Lebanon: University Press of New England.

Hillier, A. E., Culhane, D. P., Smith, T. E., & Tomlin, C. D. (2003). Predicting housing abandonment with the Philadelphia neighborhood information system. *Journal of Urban Affairs, 25*(1), 91–106.

Hirokawa, K. H., & Gonzalez, I. (2010). Regulating vacant property. *The Urban Lawyer, 42*(3), 627–637.

Huuhka, S. (2016). Vacant residential buildings as potential reserves: A geographical and statistical study. *Building Research & Information, 44*(8), 816–839.

Ikeda, S. (2007). Urbanizing economics. *The Review of Austrian Economics, 20*(4), 213–220.

Ikeda, S. (2010). The mirage of the efficient city. In S. A. Goldsmith & L. Elizabeth (Eds.), *What we see. Advancing the observations of Jane Jacobs* (pp. 24–33). Oakland: New Island Press.

Ikeda, S. (2017). The city cannot be a work of art. *Cosmos and Taxis, 4*(2/3), 79–86.

ISTAT. (2014). *Edifici e abitazioni*. Available at https://www.istat.it/it/files//2014/08/Nota-edifici-e-abitazioni_rev.pdf. Accessed December 2018.

Kaganova, O., & Nayyar-Stone, R. (2000). Municipal real property asset management: An overview of world experience, trends and financial implications. *Journal of Real Estate Portfolio Management, 6*(4), 307–326.

Kraut, D. T. (1999). Hanging out the no vacancy sign: Eliminating the blight of vacant buildings from urban areas. *New York University Law Review, 74*, 1139–1177.

Mallach, A. (2006). *Bringing buildings back: From abandoned properties to community assets*. New Jersey: Rutgers University Press.

Mangialardo, A., & Micelli, E. (2016). Social capital and public policies for commons: Bottom up processes in public real estate property valorization. *Procedia Social and Behavioral Sciences, 223*, 175–180.

Mangialardo, A., & Micelli, E. (2018). From sources of financial value to commons: Emerging policies for enhancing public real-estate assets in Italy. *Papers in Regional Science, 97*(4), 1397–1408.

Marona, B., & van den Beemt-Tjeerdsma, A. (2018). Impact of public management approaches on municipal real estate management in Poland and The Netherlands. *Sustainability, 10*(11), 1–15.

MEF (Ministero dell'Economia e delle Finanze). (2015). *Rapporto sui beni immobili delle Amministrazioni Pubbliche*. Available at http://www.dt.tesoro.it/export/sites/sitodt/modules/documenti_it/patrimonio_pubblico/patrimonio_pa/RapportoImmobili_DatiAnno2015.pdf. Accessed December 2018.

Micelli, E., & Mangialardo, A. (2016). Riuso urbano e immobili pubblici: la valorizzazione del patrimonio bottom up. *Territorio, 79*, 109–117.

Morckel, V. (2014). Predicting abandoned housing: Does the operational definition of abandonment matter? *Community Development, 45*(2), 121–133.

Moroni, S. (2012). Why nomocracy: Structural ignorance, radical pluralism and the role of relational rules. *Progress in Planning, 77*(2), 46–59.

Moroni, S. (2015a). Complexity and the inherent limits of explanation and prediction. *Planning Theory, 14*(3), 248–267.

Moroni, S. (2015b). *Libertà e innovazione nella città sostenibile. Ridurre lo spreco di energie umane*. Roma: Carocci.

Moroni, S. (2018). Property as a human right and property as a special title. Rediscussing private ownership of land. *Land Use Policy, 70,* 273–280.
Moroni, S., & Minola, L. (2019). Unnatural sprawl: Reconsidering public responsibility for suburban development in Italy, and the desirability and possibility of changing the rules of the game. *Land Use Policy, 86,* 104–112.
Olivadese, R., Remøy, H., Berizzi, C., & Hobma, F. (2017). Reuse into housing: Italian and Dutch regulatory effects. *Property Management, 35*(2), 165–180.
Palmberg, J. (2013). Spontaneous orders and the emrgence of economically powerful cities. *Cosmos and Taxis, 1*(1), 21–29.
Rawdanowicz, L., Wurzel, E., & Ollivaud, P. (2011). *Current issues in managing government debt and assets*. OECD Economics Department Working Papers, No. 923, OECD Publishing.
Shoked, N. (2014). The duty to maintain. *Duke Law Journal, 64,* 437–513.
Tanzi, M. V., & Prakash, T. (2000). *The cost of government and the misuse of public assets*. International Monetary Fund working paper. IMF: Washington.
Trojanek, M. (2015). Strategic municipal real estate management. *Journal of International Studies, 8*(2), 9–17.
Van Doren, P. (2005). The political economy of urban design standards. In E. Ben Joseph & T. S. Szold (Eds.), *Regulating place*. London: Routledge.
Vermiglio, C. (2011). Public property management in Italian municipalities: Framework, current issues and viable solutions. *Property Management, 29*(5), 423–442.
Vincent, J. (2012). Neighborhood revitalization and new life: A land value taxation approach. *American Journal of Economics and Sociology, 71*(4), 1073–1094.
Vitale, T. (2009). Socialità, mobilitazione e innovazione sociale nelle città europee. In G. Anomale (Ed.), *Il dire e il fare: volontari creativi per il bene comune* (pp. 11–21). Bine: Milano.

Chapter 3
Participation, Culture, Entrepreneurship: Using Public Real Estate Assets to Create New Urban Regeneration Models

Alessia Mangialardo and Ezio Micelli

Abstract For years, public real estate assets were considered a reserve available to compensate limited financial resources of the central government. When demand ceased to exist, many public buildings were underused or abandoned. Sometimes with illegality forms of employment tolerated by the owner administrations, sometimes with the more or less support of public authorities, communities of various kinds have found in the public assets the place to create new forms of social and economic organization offering hospitality to new formations of urban life, associations and social entrepreneurs capable of economic and social innovation whose collective benefits concern the city as a whole. The paper aims to address the issue by focusing attention to many changes within the Italian context that, from a technical and cultural point of view, appear important in the broader reasoning on the forms of urban regeneration and policies to support it. Although the processes of enhancement of public real estate assets are a well-established procedure, the intrinsic fragility of the phenomenon requires adequate policies, certainly endowed with financial resources, but above all capable of a renewed cultural attitude.

3.1 Introduction

From a reserve of financial value to a resource that enables a new season of economic and social development of cities: the non-strategic public real estate assets have ceased to be the exclusive object of market development and have begun to host activities of horizontal subsidiarity on various fronts, from innovative welfare to digital manufacturing, from performing arts to urban agriculture with an important role in the processes of urban regeneration.

A. Mangialardo (✉) · E. Micelli
Department of Architecture and Arts, IUAV University of Venice, Dorsoduro, 2206, 30123 Venice, Italy
e-mail: amangialardo@iuav.it

E. Micelli
e-mail: micelli@iuav.it

© Springer Nature Switzerland AG 2020
I. M. Lami (ed.), *Abandoned Buildings in Contemporary Cities: Smart Conditions for Actions*, Smart Innovation, Systems and Technologies 168,
https://doi.org/10.1007/978-3-030-35550-0_3

Therefore, the value of the public real estate assets changes: it is no longer just the financial value, linked to improbable alienation to developers and investors, that is often impossible to achieve—especially outside the large metropolitan areas—but it offers hospitality to associations and social entrepreneurs capable of economic and social innovation whose collective benefits concern the city as a whole.

However, the current change raises a number of issues that intersect different disciplines and areas of public administration that are not used to confrontation and cooperation. The chapter aims to address the issue by focusing attention to many changes in the Italian context that, from a technical and cultural point of view, contribute to the debate on the forms of urban regeneration and policies to support it.

The chapter is divided into five points. The first concerns the new dimensions of the value of public real estate assets. In the second section the forms of participation and their evolution are examined. The third point considers the centrality of production by the emerging figure of the social entrepreneur in the regeneration of heritage and the city. The fourth considers the role of culture in the regeneration processes promoted by the grass-roots participation. The last point outlines the implications of urban policies and focuses attention on their no longer regulatory but enabling nature.

3.2 The Assets Involved

For years, real estate assets were considered to be a valuable reserve available to governments, useful to compensate limited financial resources of the central government. The national legislator has contributed with specific procedures in this regard and has allowed an increasingly easy conversion of the real estate value, relating to assets no longer considered strategic and functional to the services promoted by the owners, into financial value.

Among all, the rule contained in the Italian Finance Act of 2008 is exemplary: the municipal administration, interested in enhancing its real estate assets, has considerably simplified the procedures for disposing of real estate no longer useful for public purposes. In particular, the asset to be disposed can be placed on the market with a new urban destination through a single procedure in order to maximize its value.

A similar emphasis on the supply side has been for years the main measures to enhance the value of disused public real estate assets, making its use sub-optimal. In this sense, the objective of the public policies was to transfer the asset to the market operators in the hypothesis of returning an enhanced stock in order to benefit the operators themselves and, more generally, the whole community.

The breakthrough that has taken place in the market since the end of the last decade has changed the overall framework where the traditional logic of asset enhancement has taken place. For years, demand has been considered always available for the enhancement and development of demanding assets.

According to the Italian context, we highlight two numbers among all: the titles enabling the construction—building permits, a good proxy for the intensity of real estate development—of the real estate market have gone from 306 thousand in 2005 to 58 thousand estimated by ANCE in 2013, with the effect of 6 million assets abandoned or underutilized (Campagnoli 2014). The consequence has been the disappearance of investors and developers for years always attentive to the disposals promoted by the public sector.

The change, therefore, does not seem to be due to a temporary economic situation outdated to return to a glorious market as that of the first half of the last decade. The forms of development of the country—basically concentrated in a few metropolitan areas—and its demography does not allow to be optimistic about the possibility of a rapid recovery of market prices and quantities for large portions of the national territory.

When demand ceased to exist, many assets remained in the hands of the owner's administrations without effective use, therefore, consequently they were abandoned or underutilized. A vast and differentiated building stock—just think of the assets intended for defense (Antoniucci and Marella 2018; Artioli 2016)—has thus been the subject of new attention from those interested in finding spaces to create processes of self-organization and active participation. Sometimes in the forms of illegality tolerated by the owner administrations, sometimes with the more or less convinced support of the municipalities, communities of various kinds have found in the public assets the place to create new forms of social and economic organization (Bellè 2018; Bottero et al. 2019; Micelli and Mangialardo 2016; Mangialardo and Micelli 2018). The result was to reveal to the same administrations that the forms of value were broader than those relating to the simple transfer of assets from the public to the private sector: the public real estate has been transformed into a resource available to the community—in common, following an expression not without ambiguity (Maddalena 2014; Mattei 2011; Moroni 2015)—able to support the action of branches in the most diverse areas of the country.

3.3 The Forms of Participation

The participation promoted within such experiences deserves to be usefully deepened to emphasize its originality. In these experiences, in fact, participation is characterized by two aspects: being selective and focused on production.

For a long time, we have imagined participation as a process capable of integrating the mechanisms of political demand, that the parties and the traditional representation forces could not effectively transmit to the places of collective decision. Participation has been, in many cases, functional to represent a completer and more articulated picture of the instances compared to what the subjects traditionally assigned to this purpose are able to do (Sennett 2018).

The processes of occupation and reuse of public real estate assets do not have the ambition to bring to attention unresolved dimensions of collective demand but concern the ability of some groups to promote projects in various fields on the basis of skills otherwise inhibited and unable to promote and to represent themselves.

The Atelier for the repair and construction of bicycles and other activities at the former Piave barracks in Treviso, the activities of performers and the production of contemporary art at the CasermArchelogica in San Sepolcro—to cite a few examples—share the centrality of the activity of production, of the undertaking of those who would otherwise have no space and opportunity to express their ability to give life to goods and services of an innovative nature (Campagnoli 2014).

These processes certainly concern the active participation of the local citizenry and the ability to create paths of self-organization that otherwise are impossible or severely limited.

It would also be wrong to consider the availability of public real estate assets as the mere availability of physical sites for a subsidiary welfare system capable of compensating a shrinking public offer. The real estate assets of the administration host social formations that ask for space to promote their own projects. As a result of new and unprecedented relationships, new relational goods (Bruni 2012) form and support new visions of non-institutionalized cities that planning has not been able to predict and regulate (see p. 11 in Ostanel and Attili 2018). Such processes can remain unheard or, in virtuous cases, they can activate institutional learning paths capable of reconnecting self-organized groups and local authorities (see p. 8 in Ostanel and Attili 2018).

The new value of the assets appears clear in this passage: from a reserve of unlikely financial value to a set of assets that allow the community to support initiatives that the market has no interest in enhancing, especially in the less advantaged areas of the country. In this perspective, public assets do not enjoy a specific legal status, while they are characterized by the enabling support function in favor of innovative processes from an economic and social point of view (Arena and Iaione 2012).

3.4 The Rule of the Social Entrepreneur

The theme of production and enterprise is central to grasping the originality of the processes in place. We do not want to ignore the importance of other activities more traditionally linked to the integration of collective services, often lacking in places where the processes of social self-organization are based. In the experience of ExFadda in San Vito dei Normanni there is also a competitive library compared to the public one and in the already mentioned former Piave barracks in Treviso there is a reception service for migrants to integrate the local welfare. However, what unites and distinguishes the now numerous experiences of regeneration of public assets is the desire of new emerging figures to create original formulas for the production of goods and services (Cucinella 2018).

The presence of the figure of the social entrepreneur (Bailey 2012) is central to the success of the regeneration of places (Gallent and Ciaffi 2014; Mangialardo and Micelli 2017; Tricarico 2016). We can distinguish two specific situations: if a social entrepreneur stands out in the process, he will be the subject able to federate the many souls of a renewed participation and to promote the initiatives undertaken, adapting them to the mutable circumstances of the local economy and society; if this figure is not present in the process, the experiences will tend to disperse the original wealth of the initiatives undertaken. (Mangialardo 2017).

The adjective social is central to grasping the nature of the processes underway. It is not an issue of giving space to companies that seek (exclusively) to profit their natural mission by taking advantage of the opportunities of a patrimony granted free of charge, as happens in most cases. Instead, it is a matter of giving space to entrepreneurs whose mission seems to be characterized by a double value. First of all, it takes on a risk that is normally higher than that which distinguishes associations and the non-profit sector in general. Second, as a social worker, he operates in a consciously interesting way in the creation of collective value for the city and the neighborhood, unlike other subjects such as cooperatives (see p. 628 in Ragozino 2019).

In the many experiences of the country, which can be traced back to a new model of "hybrid enterprises" (Venturi and Zandonai 2016), the node of the creation of social value is evident: doing business in the cultural, advanced manufacturing and agricultural fields is functional to the construction of new links with the city and the territory; it represents the element of reconstruction of ties otherwise destined to dissolve—as in particular in the case of the experiences promoted in the southern regions and internal areas (Cerreta et al. 2019; Venturi and Zandonai 2019; Mangialardo and Micelli 2016).

Production and enterprise, therefore, assume their own sustainability not only and not so much as an economic fact in itself, but as an element capable to organize, in an original way, a human capital and a social capital otherwise destined to be often irretrievably lost.

Such a role of production is not only a particular aspect of the many experiences of urban and territorial regeneration in our country, but it is also confirmed by numerous international events. Of all the experiences, the Paris experience seems to us to be the most capable of summarising the change underway in a slogan *Fabriquer en ville, fabriquer la ville*: production and enterprise are central to the French capital not only to create new production paradigms but to think and plan new and stronger bonds of a society struggling with the necessary paths of collective reorganization.

3.5 The Value of Culture

Culture is constantly at the center of the action of the social entrepreneurs who regenerate the "out-of-the-ordinary spaces" (Ostanel 2017) of the public assets that have been taken away from abandonment. The attention devoted to a concept with uncertain borders appears peculiar to our country on the international scene and is potentially one of the driving factors for a new path of growth (Caliandro and Sacco 2011; Forte and De Paola 2019). However, even in this case, it seems necessary to specify how the concept of culture finds space in processes that, as we have argued, find their distinctive feature in the production of goods and services.

Culture assign a new meaning to the production of goods and services and, consequently, helps to restore abandoned spaces. Let's return to the case of ExFadda, which appears enlightening to explain the connections just outlined. In San Vito dei Normanni a group of young people has started to produce blankets made from waste wool. A product of the local tradition aimed at the reuse of what would otherwise be wasted. The legacy of tradition is, however, reinterpreted through a new design, production cycles that go beyond domestic production, without neglecting various aspects of marketing—especially with regard to the brand of new products—which represent an aspect that unites many of the projects of ExFadda.

The project concerns traditional activities extensively revised through contemporary production and marketing logic. The synthesis of traditional manufacturing and contemporary knowledge is the basis of the business project, as it happens in many national productions where innovation is not necessarily linked to products with a high technological content, but rather to the ability to immerse them in the historical context of a local scenario or to attribute to them a cultural value linked to the material processes of their production (Micelli 2016).

If, however, the cultural content transforms tradition and brings it into the contemporary, what is striking is the community bond at the base of the new production: old ladies share their knowledge and bring it into play with young designers in a project that finds in the strengthened generational bond a qualifying point confirming that culture not only creates economic value but contributes to the production of new social bonds (Sacco et al. 2012).

Other examples highlight the regenerative possibilities of knowledge and culture: agriculture in abandoned areas regains momentum as a result of investment in new crops, personal services change as a result of investment in new forms of body culture. Traditional manufacturing is innervated by new cultures—sometimes antagonistic—of makers and digital craftsmen. If the culture and knowledge of a generation do not find a place in the market and in the public sector, then the human capital of the youngest (Campagnoli 2014) occupies the spaces of a society that has mistakenly believed in the immutable value of real estate (Dalla Zuanna and Weber 2012) to test the hypothesis that the creation of value is based on the cultural and symbolic wealth of goods and services and does not necessarily concern only the company, but also the community.

3.6 Enabling Policies Versus Participation Policies

The success of bottom-up regeneration experiences is recognized in events and publications. It is not only the academic and scientific culture or the major cultural events that deal with it. The protagonists of this season of development of bottom up processes are also several foundations that have sensed the extent of the phenomenon and have promoted its development. Among other things, the example of the Unipolis Foundation's Culturability[1] call for proposals certainly deserves the attention of the case. There are now hundreds of associations and groups participating in the call, on the one hand, to obtain a financial contribution that, although not particularly important if compared to the needs of transformation of the assets at stake, is still not negligible; on the other hand, the success of the call promoted by the foundation in Bologna allows an important accreditation with public and private institutions in view of further funding and other forms of economic support.

However, it is difficult to take off a public policy in support of such initiatives. The reasons for the lack of precise policies in this direction can be attributed in the first instance to the modesty of financial resources available to local authorities, which are the first institutional actors to come into contact with self-organized training. Moreover, the administrators have already struggled to use the existing juridical institutions to favor the entry of the new subjects in assets otherwise destined to ambitious development programs, economically unrealistic and solid in their administrative foundations (Mangialardo and Micelli 2017).

However, the scarce financial resources represent an obstacle that hides others of cultural nature. In a nutshell, when the administrations accept the perspective of the training development, based on a social enterprise project often with uncertain and mutable borders, they change their actions to the roots. Indeed, they transform themselves from subjects that envisage the development of public assets and indicate precisely how it is to be implemented (this is the case, for example, of urban planning) into subjects that enable change in the knowledge that they have triggered processes and initiatives with uncertain and unpredictable results. Uncertain because it is not taken for granted that the social entrepreneur will eventually achieve the result of balancing the creation of value for himself and for the community in contexts of modest demand for quantity and availability of expenditure; unpredictable because the undertaken can change direction and change over time as a result of defections, new alliances, and new competitors. Generative planning (Carta 2017) does not foresee, does not pursue any final state of the places (Moroni 2007). On the contrary, the aim of the social entrepreneur is to maximize the opportunities for expression of groups and associations in the perspective of an open and plural society.

The enabling policies are completely different from the sectors that are traditionally called to govern the development of the territory (Cottino and Zeppetella 2009). They have different assumptions and impose innovative behavior over time compared to the past: it is a matter of monitoring, accompanying and evaluating changes in the course of development in the light of renewed boundary conditions. Policies of this nature require not only adequate resources but also a different cultural attitude and operational tools compared to the past. And both elements—dedicated funding lines and that of profoundly renewed policies and lines of action—represent demanding challenges, the success of which appears essential not only to give continuity and solidity to the current season but also to prefigure planning capable of new objectives and new instruments.

[1]To know in detail this project please refer to the official website: http://www.fondazioneunipolis.org/progetti/culturability/.

3.7 Conclusions

Public real estate assets change their role in the urban agenda. It ceases to be a simple reserve of financial value to compensate the economic difficulties of administrations and it is transformed into a resource capable of hosting social and economic innovation. The phenomenon, which started from the grassroots participation and was subsequently adopted, still only partially, by the administrations, crosses several themes: it transforms participation by changing its constituent aspects, it makes crucial the processes of production of goods and services, enhancing their positive externalities for the benefit of the community, it assumes culture and knowledge as inescapable factors that generate value.

However, the phenomenon shows clear signs of fragility. If the scientific literature and the numerous initiatives on the topic exalt the successful experiences, strengthened by their independence and their capacity for self-organization, it would, however, be wrong to consider public support irrelevant. The intrinsic fragility of the phenomenon requires adequate policies, certainly endowed with financial resources, but above all capable of a renewed cultural attitude, attentive to promoting processes whose development and the outcome are plural and substantially indeterminate. The transition to enabling policies, capable of accepting the intrinsic uncertainty of urban change, is not only the most important challenge of collective action in the new forms of enhancement of public assets but is a challenge for the whole culture of the government of the territory.

References

Antoniucci, V., & Marella, G. (2018). Is social polarization related to urban density? Evidence from the Italian housing market. *Landscape and Urban Planning, 177,* 340–349.

Arena, G., & Iaione, C. (2012). *L'Italia dei beni comuni.* Roma: Carocci.

Artioli, F. (2016). Le aree militari nelle città italiane: patrimonio pubblico e rendita urbana nell'era dall'austerity e della crisi. *Rivista delle Politiche Sociali—Italian Journal of Social Policy, 1,* 89–113.

Bellè, B. (2018). Iniziative bottom-up e riuso temporaneo. Quale valore aggiunto per la valorizzazione di beni immobili pubblici? In *CRIOS* (vol. 16, pp. 35–44).

Bailey, N. (2012). The role, organization and contribution of community enterprise to urban regeneration policy in the UK. *Progress in Planning, 77*(1), 1–35.

Bottero, M., Oppio, A., Bonardo, M., & Quaglia, G. (2019). Hybrid evaluation approaches for urban regeneration processes of landfills and industrial sites: the case of the Kwun Tong area in Hong Kong. *Land Use Policy, 82,* 585–594.

Bruni, L. (2012). *The wound and the blessing. Happiness, economics, relationships.* New York: New City Press.

Carta, M. (2017). *Augmented City.* List, Rovereto: A Paradigm Shift.

Cottino P., & Zeppetella P. (2009). *Creatività, sfera pubblica e riuso sociale degli spazi.* Paper 4, Cittalia, Roma http://www.osservatorioriuso.it/cgi-bin/documentazione/Paper4-09_Cottino_Zeppetella.pdf.

Caliandro, C., & Sacco, P. L. (2011). *Italia reloaded.* Il Mulino, Bologna: Ripartire con la cultura.

Campagnoli, G. (2014). *Riusiamo l'Italia. Da spazi vuoti a start-up culturali e sociali*. Gruppo 24 Ore, Milano.

Cerreta M., Poli G., Regalbuto S., & Mazzarella C. (2019). A multi-dimensional decision-making process for regenerative landscapes: A new harbour for Naples (Italy). In S. Misra et al. (Eds.), *Computational science and its applications—ICCSA 2019*, Lecture Notes in Computer Science, vol. 11622. Cham: Springer.

Cucinella, M. (a cura). *Arcipelago Italia. Progetti per il futuro dei territori interni del Paese. Padiglione Italia alla Biennale Architettura 2018*, Quodlibet, Macerata.

Dalla Zuanna G., & Weber, G. (2012). *Cose da non credere. Il senso comune alla prova dei numeri*. Bari: Laterza.

Forte, F., & De Paola, P. (2019). How can street art have economic value? *Sustainability, 11*, 580.

Gallent, N., & Ciaffi, D. (Eds.). (2014). *Community action and planning: Contexts, drivers and outcomes*. Bristol: Policy Press.

Maddalena, P. (2014). Il territorio bene comune degli italiani. Proprietà collettiva, proprietà privata e interesse pubblico. Roma, Donzelli.

Mangialardo, A. (2017). Il social entrepreneur per la valorizzazione del patrimonio immobiliare pubblico. *Scienze regionali, 16*(3), 473–480.

Mangialardo, A., & Micelli, E. (2016). Social capital and public policies for commons: Bottom up processes in public real estate property valorization. *Procedia—Social and Behavioral Sciences, 223*, 175–180.

Mangialardo, A., & Micelli, E. (2017). La partecipazione crea valore? Modelli di simulazione per la valorizzazione dal basso del patrimonio immobiliare pubblico. Valori e Valutazioni, 19, 41–52.

Mangialardo, A., & Micelli, E. (2018). From sources of financial value to commons: Emerging policies for enhancing public real-estate assets in Italy. *Papers in Regional Science, 97*(4), 1397–1408.

Mattei, U. (2011). *Beni Comuni. Un manifesto*. Roma-Bari: Laterza.

Micelli, E., & Mangialardo, A. (2016). Riuso urbano e immobili pubblici: la valorizzazione del patrimonio bottom up. *Territorio, 9*, 109–117.

Micelli, S. (2016). *Fare è innovare*. Bologna: Il Mulino.

Moroni, S. (2007). *La città del liberalismo attivo*. Milano: Cittastudi.

Moroni, S. (2015). Beni di nessuno, beni di alcuni, beni di tutti: Note critiche sull'incerto paradigma dei beni comuni. *Scienze regionali, 14*(3), 137–144.

Ostanel, E. (2017). *Spazi fuori dal Comune: Rigenerare, includere, innovare*. Milano: Angeli.

Ostanel, E., & Attili, G. (2018). Self-organization practices in cities: Discussione the transformative potential. *Tracce Urbane, 4*, dicembre, 6–17. http://ojs.uniroma1.it/index.php/TU.

Ragozino, S. (2019). Navigating neo-liberal urbanism in the UK. Could a social entrepreneur be considered an activist planner? In F. Calabrò, L. Dalla Spina, & C. Bevilacqua (Eds.), *New metropolitan perspectives* (pp. 625–634). Cham: Springer.

Sacco, P. L., Ferilli, G., & Tavano, Blessi G. (2012). Sviluppo locale a base culturale: quando funziona e perché? Alla ricerca di un framework di riferimento, in Prisma. *Economia, Società, Lavoro, 1*, 9–27. https://doi.org/10.3280/PRI2012-001003.

Sennett, R. (2018). *Costruire e abitare. Etica per la città*. Milano: Feltrinelli.

Tricarico, L. (2016). Imprese di comunità come fattore territoriale: riflessioni a partire dal contesto italiano. In *CRIOS* (vol. 11, pp. 35–50).

Venturi, P., & Zandonai, F. (2016). *Imprese ibride. Modelli d'innovazione sociale per rigenerare valore*. Milano: EGEA.

Venturi, P., & Zandonai, F. (2019). *Dove. La dimensione di luogo che ricompone impresa e società*. Milano: EGEA.

Chapter 4
Intensity of Uses and Spatial Devices

Lucia Baima and Matteo Robiglio

Abstract Through the concept of intensity, this chapter intends to propose a different point of view to approach new questions posed by the contemporary urban context, concerning the reactivation and recoding of uses and practices within consolidated city spaces. Today, just as during the "spiritual crisis" of the '70s, free plots, abandoned buildings, empty apartments, rooms, roofs, and all the city surfaces become magnets that are able to create additional capacity for the unspoken desires, as able to the visible and invisible characteristics of spaces. In other words, thanks to different practices and phenomena, spaces are re-thought (or re-conceived) as a platform for different possibilities over time. Intensity, rather than density, is a tool capable of showing this potential in a kinetic process. Through an etymological analysis, the interpretation of the role of the project in generating vitality within the urban space. Hereafter, the intensity acts as a key to reveal the potential of space to catalyze un-programmed and free uses not yet conceived thanks to its design and devices. In other words, intensity allow conceving project as a process. All the spaces become a platform for interaction and opportunity—from the house to the city.

4.1 Space as a Receptor for Action!

Recently urban studies describe cities as "topologies of many intersecting entities, networks, and flows whose combined force shapes urban development" (see p. 243 in Amin 2015).

Spaces catalyze this process. They act as platforms of plural geographies of association that create different kinds of spaces.

L. Baima (✉) · M. Robiglio
Dipartimento di Architettura e Design (DAD), Politecnico di Torino, Viale Mattioli 39, 10122 Turin, Italy
e-mail: lucia.baima@polito.it

M. Robiglio
e-mail: matteo.robiglio@polito.it

© Springer Nature Switzerland AG 2020
I. M. Lami (ed.), *Abandoned Buildings in Contemporary Cities: Smart Conditions for Actions*, Smart Innovation, Systems and Technologies 168,
https://doi.org/10.1007/978-3-030-35550-0_4

Amin states that "multiple logics and recombinant dynamics practice generate urban outcomes that are at one and the same time recursive and emergent, regulated and unplanned, stable and changing" (see p. 243 in Amin 2015). According to this approach, time and space logics converge and become juxtapositional spaces (Massey 2005) constantly fluctuating between order and disorder, cohabitation and conflict, continuity and change. In other words, what Lefebvre defined as balance and imbalance of the city (Lefebvre 1974). Space becomes a superposition of actions over time that contributes defining cities as fields of transformation.

Transformation of space-time uses with multiple—sometimes paradoxical—practices remarks that space is a social product constantly under construction (Lefebvre 1974; Harvey 1990; Bauman 2000; Massey 2005).

The recent 2008 real estate crisis encouraged the emergence and development of practices centered on sharing resources—economics, physicals, interests—and above all shared spaces—often starting from un-used or sub-used built infrastructure. Reframed as an available and/or shared resource, a single space becomes many places for temporary access to multiple services (Botsman and Rogers 2010).

Space becomes therefore an experimental laboratory for these new forms, uses and practices of urban living as well as for alternative urban lifestyles (see p. 3 in Zukin 1982). Space enables and intensifies their relationship with and in the city. These processes and practices can be described through the re- and co- suffixes: re-use, re-generation, re-adapting as well as co-production, co-operation, etc. They conceive space as their platform and enabler that reveal its potential throughout the reprogramming of use over time (Oswalt et al. 2013).

From the micro to the macro scale, these practices emerge from a constant redefinition of the boundaries between public and private space and the programmed functions that make room for other unexpected ones.

Temporary and intermittent formulas of habitation emerge from the private dimension of the house, which generates an elastic contraction of the more private space, which is temporarily transferred to functions that recall a more shared and public dimension of urban living.

At the same time, free plots, terrain vague, empty apartments and buildings, and all the related surfaces can become catalysts of different uses and practices. As Saskia Sassen says, these spaces are "potentiality very productive spaces not just in terms of informal economies but also in terms of making temporary public spaces" (see p. 109 in Sassen 2013) and temporary uses "these (…) include eventization of urban spaces and as well as the spatio-temporal dynamization of services" (see p. 10 in Oswalt et al. 2013).

The urban public spaces become a relational and performative platform just like a table "is located between those who sit around it (…) like every *in-between*[1] relates and separates different (actors) at the same time" (see p. 52 in Arent 1958).

[1] "To live together in the world means essentially that a world of things is between those who have it in common, as a table is located between those who sit around it; the world, like every in between, relates and separates men at the same time. The public realm, as the common world, gathers us together and yet prevents our falling over each other, so to speak". Arendt (1958:1998), *The human condition*. Chicago: The University of Chicago Press, p. 52.

In doing so, all the city surfaces are ready to accomodate multiple uses in multiple times and can become a place—or infinite places—for an encounter between individuals and space, in which they see a potential of something that is already there. This process of transformation of different active intensity generates uncertain effects, in experimental spaces: hybrid, indeterminate, heterogeneous, temporary, unstable, reversible. Such experimental spaces value increased by the presence of temporary inhabitants. Space is thus discovered as a support for heterogeneous actions in a time considered of multiple dimension, non-linear, but simultaneous as "a network that connects points and intersects with its tangle" (see p. 19 in Foucault 1967).

Within this time frame, the spaces are activated and redefined by different and multiple actions, practices, performances, or types of public.

These are alternately singles/plurals, privates/public, void/full, charged/empty, suspended/continuous, off/intense; the uses and activities are similar/different, divided/mixed, alternate/coexistent.

This dynamic of rewriting the cities is generated by a complex panorama of actions.[2] Actions take place everywhere within the city, and their development is incremental during the time.

They show and expand the idea of the space potential, which is encouraged by availability and where alternatives can be found.[3]

This process is able to transform everyday, a space in a place (Heidegger 1954).

The place is where someone decides to place or position themselves, in other words, it is where the space acquires meaning for an individual (Tuan 1977). As well as, the place is the time when people add to space, unexpected and uninterrupted possibilities—once they made it their own (Hertzberger 1991).

Thus, multiple possibilities of use are defined for variable temporalities that outline a web of potential places in which to activate and reactivate the process of action (Hertzberger 1991). Therefore, the concept of inhabit/living the city becomes wider.

---

[2]"This polymorphous dynamic, clandestine or official, militant or more institutional, informal or more organized, often arises in industrial wastelands, public or abandoned spaces where new, responsible ways of building, living or sharing can be tested out.

(…) In these open places, "Third Place activities" are the norm, "alternative buildings" are the rule, "reversibility" is a standpoint, "fragility" is an observation and "workshopping" a way of doing things that implies movement, possibility, and openness. Something like the possibility of a town begins to take shape between do-it-yourself and open innovation, urbanism and activism. (…) The blurring of time, activity, and status, the blurring of living space, scale, and recomposition, the tendency toward multipopulated and multiscalar alliances and recompositions: all these lead to a hybridization of structures, objects, and practices".

Luc Gwiaździński, *Localizing the in-finite*, in Infinite Places, Paris: Editions B42, p. 43.

[3]"(…) le possible est une tentation que le réel finit toujours par accepter", Bachelard (1932). *L'intuition de l'instant*, Paris: Stock, 45.

4.2 Raw Space, the Breeding Ground of Ideas

These dynamic processes and practices are not new manifestations. They have existed and grown since the second half of the nineteenth century.

If we analyze New York during the *spiritual crisis* (Mahler 2005) of the 70s, a city emerges under the surface. This invisible city is defined by practices that spontaneously and transversely trigger new and surprising actions of re-appropriation of the vacant spaces created by the crisis.

These are multiple processes that take place through collective involvement in the re-appropriation of the space. Its rediscovery is activated through the redefinition of the architectural and urban devices that, re-thought and re-used, become surfaces on which the discomfort is physically transferred, and the resilient reaction to the crisis is implemented.

The city rediscovers itself as an open-air laboratory.

Abandoned buildings, empty plots and all the vertical and horizontal surfaces—overall empty both physically and of interest—evolve from being detractors to resource spaces able to be reactivated.

Roofs, facades, walls, fences, as well the subway wagons surfaces, overcome the function of separation device between inside-outside and are rethought by artists and writers as supports for site-specific works that provocatively redefine their perception and detect their new use.

Thus, the potential contained within all the spaces and in every single component emerges.

The empty interstitial blocks of the grid, vacant after the demolition of foreclosed houses, are transformed in self-managed urban gardens—the community garden. These plots become spaces around which the community identifies itself. This is due to a reprogrammed use of space over time that transforms them into urban platforms with multiple intensity uses.

At the same time, abandoned buildings become subjects for collective or hybrid housing ideas: squatting and loft movement. Both of these practices arise from movements of illegal occupations by groups of inhabitants and collectives of artists who share the interior space according to a logic of maximizations of uses in space-time. This strategy is implemented thanks to minimal homemade devices and taking advantage of the potential offered by the different types of buildings—tenement and loft building.

In the first case, the spatial layout is redesigned as well as the hierarchy and the proportions between public space and private space, through home-made devices—according to the do-it-yourself logic.

The private space is contracted both to maximize the number of residential units and strengthen the relative sense of community and to expand the space for collective functions.

This strategy crosses the entire building from the basement to the roof.

The basement and the backyard are redesigned as laboratories, collective workshops, spaces for concerts and events, as well as the staircase become the real vertical common space that extends up to the rooftop that is in turn transformed into a garden and a public square at altitude.

In the second case, simple mobile devices, clinging to the original structure—sliding partitions, tip-up equipped walls, curtains, dividers, carriages, ladders, and mezzanines, etc.—are used as interchangeable cards able to create new configurations every time, to hierarchize the interior space on multiple levels, to define new functions.

The loft buildings designed for light manufacturing (see p. 13 in Zukin 1982) are thus rethought as real high-altitude creative urban laboratories for mixed-use and co-existing uses. The structure itself allows for this process thanks to its generous proportions: one thousand feet of space on each floor, a high ceiling supported by large spans between the cast-iron columns.

The features of these raw spaces release the potential or different uses follow the "drama in everyday life" (see p. 2 in Zukin 1982).

In other words, these spaces are no longer thought to accommodate a single function. They are rather intended for multiple purposes.

The aim of this architecture is to increase the highest number of dynamic experimental possibilities—it cannot stand still; it is always adapting because beginnings have no end. Therefore, space offers further degrees of possible uses which also influences the artistic process.

For instance, "if you don't have lot of room your idea get very small" (Life 1971). Creating, exhibiting and living follow each other without interruption within the same perimeter which thus accommodates exponentially hybrid and flexible functions, sometimes even irrational. The exploitable space becomes an infinite space, contracts or dilates following the unpredictability that the creative process underlies.

At the same time, the raw and crummy nature is a matrix and constant source of inspiration for the entire city. The artists indeed not only work within but on the space itself: floor, walls, ceilings, and overall architectural features.

In other words, the artists integrate the architectural and city features into their work, and the formal difference between artistic container and artistic contents are erased. Space is entirely absorbed into an artist work, and content and container are linked then, by a continuing process, to the point that "context as content" (O'Doherty 1986).

All these vacant spaces are rethought as an *alternative space* (O'Doherty 1986) that represent the desire to break out of the frame, to extend the boundaries and definition of what was considered the right use of spaces. These alternative spaces play a key role to rethink the city and its spaces. In fact, looking for cheap and vacant spaces, the artists set up their studios in any building they find, thus rethinking the original purpose of the industrial spaces they occupy and reveal the space inside space or its potential. At the same time, these spaces can changes over time, they permit to act, welcoming the unexpected, and their complexity is never fixed in its habits.

They offer everyone the possibility of action while realizing their inner potential.

> All it was quite spontaneous on purpose – no particular schedule – all is wrapped an atmosphere of immediacy. [...] the quality and spirit of the place were in this spontaneity. (Solomon 1967)

Reread overall—even if in a diachronic way—these processes impose to overcome the idea that precise identification of the space with its programmed function is always possible, but that the use of the same is the result of a temporary balance re-enacted continuously in the time.

In these experiences ingenious devices trigger ways and opportunities to spatialize serendipity. By experimenting, adjusting, opening up possibilities and trying them out, these processes expose the hybrid nature of places, their provisional character; overall their potential.

Therefore, the potential is the implicit dimension expressed through processes, which include its individual and subjective redefinition—their own way of living in it—of distributing it, configuring and reconfiguring functions, hybridizing them with different uses, creating changeable filters that they define a personal separation or conjugation between private and public sphere.

These features depend on the idea to let an active-intensity space able to the mobilization of latent resources—able to do appear what is not expect—as a workshop of potential as well as a workshop of possible.

Architecture thus becomes a "tactile appropriation, as constantly being transformed by use, its boundaries renegotiated by habits" (see p. 49 in Amin and Thrift 2002).

4.3 Intensity Instead of Density

This dynamic realm shows how necessary it is to adopt a point of view enabling the understanding of this complexity and differentiation, these overwritten practices, to redefine the terms—or at least to adopt a more sensitive vocabulary and strategies used in describing these processes (Williams 1976; Gausa et al. 2003) and to propose analytical and design tools capable of revealing their relative effects.

These three steps are necessary reactions that are able to fill the gaps between statements and real practice or as de Certeau says, "the networks of these moving, intersecting writings compose a manifold story that has neither author nor spectator, shaped out of fragments of trajectories and alternations of spaces, in relation to representation, it remains daily and indefinitely other" (see p. 93 in De Certeau 1984).

The first step is to shift the point of view: from *zenithal*—for the city seen though as a static urban area or be analyzed from above—to *oblique* (De Certeau 1980), from *synoptic to synchronic* (Harvey 1990; Sennett 2018).

Therefore, it's essential to adopt a dynamic perspective view which is able to reveal the "invisible threads that are woven between one person and another" (see p. 1026 Simmel 2009).

Finally, it is necessary to *explore* more than to *observe* (Sennett 2008).

The city, analyzed from this point of view, shows how "it will be a complicated pattern, continuous and whole, yet intricate and mobile. (…) open-ended to change of function and meaning" (see p. 118, Lynch 1960), a dynamic multiple structure that is therefore related to a dynamic temporal dimension of outcome.

The city acts as space of process and interaction, as a socio-spatial system with its own internal dynamic, like a platform where the unpredictable functions are possible. As well as relations between building and action,[4] between fixtures and flows mark the spatial and temporal openness of the city (see pp. 8–9 in Amin and Thrift 2002).

Therefore space is no longer thought of as a static agglomeration but as an *animated space* (Amin 2015) result of a *complex assemblage* (Deleuze and Guattari 1980; Dovey 2012), "whose properties emerge from the interaction between parts" (Deleuze and Guattari 1980; see p. 5 in De Landa 2005), which defined *a space is a practiced place* (see p. 111 in De Certeau 1984).

To follow this process, it is necessary to redefine the terms or to adopt a more sensitive vocabulary and to intercept these dynamic actions that feed the relational and *haptic* components of space (Deleuze and Guattari 1980). The tools are able to translate the reciprocal transformations between the container—the catalyst of actions and processes—and its content that generate an alternation of space and uses, as that Foucault says a system of *counter-sites*[5] (see p. 22 in Foucault 1986).

In fact "these practices of spaces refer to a specific form of operations ("ways of operating") to "another spatiality" (an "anthropological", poetic and mythic experience of space), and to an opaque and blind mobility characteristic of the bustling city. A migrational, or metaphorical, city thus slips into the clear text of the planned and readable city." (see p. 93 in De Certeau 1984).

Tools therefore are able to transcribe this additional dimension that exceeds the definition of space in static quantitative terms—as a *density*—but also translates its thickness, its *implicit dimension* (Tonkiss 2013) in dynamic qualitative terms—as an *intensity*.

When looking at the urban scale, density is a "simple ratio of straightforward urban quantities such as the number of people, the apartment of rooms, all divided by the

[4]"As porous as this stone is the architecture. Building and action interpenetrate in the courtyards, arcades, and stairways. In everything they preserve the scope to become a theatre of new, unforeseen constellations. The stamp of the definitive is avoided. No situation appears intended forever, no figure assert its "thus and not otherwise" W. Benjamin, *One-way street and other writing*, NLB, London, 1979, Naples, p. 169.

[5]"Whereas utopias are unreal, fantastic, and perfected spaces, heterotopias in Foucault's conception are real places that exist like "counter-sites", simultaneously representing, contesting, and inverting all other conventional sites. The heterotopia presents a juxtapositional, relational space, a site that represents incompatible spaces and reveals paradoxes". Foucault, Michel, *Of Other Spaces, Heterotopias*, Diacritics No. 16, 22–27, 1986.

geographic area over which those ingredients occur. It describes the raw ingredients that planners have to handle" (see p. 22 in Stonor 2018).

For these features, density has become a tool of volumetric and spatial prefiguration, the structuring matrix of the internal organization of the city (Amphoux 2003).

Density is the "enabler of propinquity the coming together of bodies in space" (see p. 7 in Sorkin 2003) and considers the city as a *homologous* and *static*, *isotropic* and *homogeneous* agglomerate.

In fact, it allows to measure the space, in other word defined as the container, but doesn't describe the content as an implication and the multiple relationships that it triggers—the *heterogeneous intensities* (McFarlane 2015) and *heteogeneous relations* (Thrift 2007; McFarlane 2011)—as the *essence of what it feels like to be somewhere* (see p. 24 in Stonor 2018).

Intensity allows this leap in scale and takes on a different point of view.

In fact, *intensity* is able to describe the population that's participating in the public realm of an area.

The essence of intensity is the *exchange, the transaction, the relation: social, cultural, intellectual, factual.*

These features are included in its constituted elements.

The intensity, in its generic definition, expresses the power, energy, and incidence of a phenomenon. It can look at the result over time and also describes the role that each component assumes concerning the main structure.

Thanks to these characteristics intensity easily crosses different disciplines and horizons, from concepts to sensations. It overcomes the classical metaphysical image that defines the substance as something static—fixed and determined—in favor of a substance as a temporary result of a process that is always ongoing—therefore continuous.

In fact, Deleuze defined *intensity* as a concept through which to show the sensible and to translate the dynamism of the variable forces included in the differential processes—"every intensity is differential, by itself a difference" (see p. 280 in Deleuze 1994).

It thus becomes a dynamic and *metamorphic* instrument that allows the substance to be translated as a temporary result of a process in progress, *a metastable system constantly subject to temporary equilibria* (Simondon 1964) or an eternal disequilibrium and perpetual displacement (see p. 49 in Deleuze 1988).

As Spinoza says—*power is intensity* (Deleuze 1988), *difference and variability* (Deleuze 1968) *eternal subsistence* of a *connective motion* (see p. 88 in Deleuze 1988) which produces effects in extension, in fact, "intensive differences are productive" (see p. 81 in De Landa 2005).

Intensity is, therefore, a tool with multiple focal points. It looks inside the processes and thanks to these characteristics it has extended its field of application as a tool through which to describe the *dizzying urban phenomenology* and ingredient of the *metabolism of cities* (Jacobs 1961; Salat et al. 2011).

It allows us to represent the stratification of flows—the *temporalités urbaines* (Lepetit and Pumain 1993) and the practices that maximize contact and exchange opportunities that stimulate *social interaction*, such as reciprocal tensions between subjects and spaces that structure *social experience* (Goffman 1967).

It describes the outcomes that people *experience* (see p. 24 in Stonor 2018).

Intensity shifts the attention to the intrinsic relationships of density; it becomes the third dimension of the notion of density as the sensitive and perceived sphere (Amphoux 2003).

It makes clear how heterogeneous dynamic—or the *practical correlates of everyday life* (Thrift 2014)—modify and transforms the use of space/are included within the space.

These correlations are substantial ingredients of the *vitality of the city* (Jacobs 1961).

In the face of this stratified tension, the city emerges as an *open city* (Sennett 2008) that encourages the free and subjective intervention to change and redesign the space.

The space is re-conceived as an *open system* (Sennett 2008) able to add and promote such processes of improvisation, adaptations, additions. These actions establish an open relationship with the space that translates the evolution of urban time itself—*the public realm is a process*! (Sennett 2010).

Within such frame is possible to extend the concept of intensity as an architectural tool to show and to capture the complexities of the contemporary city, in other words, to transform them into a resource for a project.

The city and its spaces, reinterpreted through the intensity, become a *semi-lattice structure* (Alexander 1965): spaces become an interference field where flows, functions, and even unexpected uses are intensified. These processes redefine original and complex spatiality through the hybridization of boundaries that in turn generate as many hybrid spaces. What can be noted is that the emergence of those unexpected dichotomies is seen at the beginning between inside-outside, public-private, permanent-temporary, planned-unexpected.

Intensity reveals two levels of action: *actions on the space* and *actions on the process*. Both of the interpretation allows us to think of the space as a generator of potentiality as provided with the ability to take on different uses and meanings, and so to incorporate different uses and practices in other word, project becomes a vector of transitions.

Space is put into tension and becomes a *performative space* (Amphoux 2003) through its design and its devices. Devices work as an *attractor*, as De Landa says "while they are not possibilities, they do act as the structure of a space of possibilities" (see p. 83 in De Landa 2005).

Intensity highlights a project potential as a generator of possibilities—or different relationships—that exceed the physical dimension of the project itself—micro-macro, open-closed, private-public.

Finally, the potential of the space is encouraged by its own availability, as well as it is amplified by its design rethought as a system for increasing the space, thus creating higher intensity. This tool shows the capacity of those spaces—thanks to aligning both physical and spatial factors or through the devices—to find new and unexpected uses in all conditions, scales or typology.

Evoking the comparison between density and intensity, introduced at the beginning of the paragraph, it is possible to say: while *density* is intended as a static and isotropic dimension, *intensity* is intended as a *kinetic and anisotropic* dimension, that connects space and time.

Density and *intensity* as the same relation of *striated space* and *smooth space:* "the striated is that which intertwines fixed and variable elements, produces an order and succession of distinct forms (…).

The smooth space is the continuous variation, continuous development of form; (…) it is filled by events or haecceities, far more than by formed and perceived things. (…) It is haptic rather than optical perception. (…) It is an intensive rather than extensive space, one of distances, not a measures and properties. *Intense Spatium* instead of *Extension.* (…) Perhaps we must say that all progress is made by and in *striated space*, but all becoming occurs in *smooth space*. (see pp. 500–510 in Deleuze and Guattari 1987).

Intensity looks at the physical dimension of space. From this transversal point of view, space assumes a porous, rhythmic, and atmospheric consistency that challenges to develop new strategies, new ways, and tools to explore, describe, and attune to it.

Space becomes an intensive and kinetic process.

In this perspective, space becomes a proactive platform where actions, practices, and processes outline trajectories that intensify the uses, the relations, and experiences in space-time. At the same time, intensity becomes a parameter through which to transcribe this process thanks to the ability to relate its ingredients: space, uses, and time.

4.4 Space as a Platform of Different Possibilities

Following this frame and thanks to it inner features, we can use intensity as a tool for analyzing and surveying how the *platform* functions and works.

Through the use of the *intensity* parameter, space emerges in all its depth and richness.

Space acts as a load space of multiple processes and practices that translate urban complexity as a space-time compression that amplifies the intensity of the relationships that become multiple and complex (Giddens 1987; Harvey 1990).

Space, through its design, its devices, its physical features, and internal organization, takes on an active and central role in determining this thickness. At the same time, it is being the background, and the informing and the soliciting structure of human behavior.

Therefore, it acts both as a platform and as a frame of interaction (Goffman 1967). It is the theater of practices, between the stage—the *public space*, and the scenes—the *private one* (Goffman 1967; Sennett 2018). Designing space conditions the relationships between an individual and other subjects. Design process acts on the *wechselwirkung*—reciprocal action—through a plurality of spatial elements, that modify their relationship in a dynamic change, and that feeds the *vergesellschaftung*—sociability. Thus, this process influences the dynamic and diachronic processes, through which the invisible threads of interaction are established and maintained (Simmel 1908).

In this frame, city becomes a laboratory that is able to catalyze different intensities (Gehl 1980; Whyte 1980), and space can be rethought as a loft.[6] Loft space performs as an open structure that catalyzes the processes of flexibility, adaptability, and resilience (Baum and Christiaanse 2012).

Devices are designed as *open filter*s (Sennett 2008) able to bring out/reveal possible interstices, whereas the project is a field of action, a paradigm of the subjective appropriation because the "space is the experience of it" (see p. 8 in Ligtelijn 1999).

Architectural devices act as *vital organs*, (see p. 29 in Jacobs 1961) tiles/pieces that feed the urban dynamism and from which diversified paths and adaptations of use of space over time are branched.

This is the phase that characterizes the 60s–70s of the last century.

In these years we've moved from the project as a *functioning machine*—structured according to the four functions of daily life, as *living, working, circulation and recreation*—to project as a *gear/apparatus.*

Space is then rethought as a continuously *attuned instrument* (see p. 170 in Hertzberger 1991), starting with the city conceived as a *field of interaction* between the individual and space—from a house, passing through a street, neighborhood and city (Smithson and Smithson 1953).

It is revealed as a hierarchy of overlapping configurative systems, conceived multilaterally.

Space takes on the initiative to act as the promoter for who gets things moving (Smithson and Smithson 2005).

Therefore, an original look at space emerges, which is a field of actions, relationships, and complex associations which thus assumes an active role in catalyzing these energies—that is its potential—whether it is micro/macro, private/public, full/empty. In fact "the charged void describes architecture's capacity to charge the space around it with an energy which can join up with other energies, define the nature of things that might come, anticipate happening (…) a capacity we can feel and act on, but cannot necessarily describe or record" (see p. 11 in Smithson and Smithson 2005).

A project shifts from a systematization device to a suggestive tool that is similar to a set of instructions for city use. A structure, that therefore includes interpretation, establishes an open dialogue between inhabitants and cities punctuated by free associations that the structure itself welcomes and urges (if thought of as mat). It works as a *mat* that "can epitomize the anonymous collective; where the functions come to enrich the fabric, and the individual gains new freedoms of action through a new shuffled order, based on interconnection, close-knit patterns of association and possibilities for growth, diminution, and change" (Smithson 1974).

[6]"We use the word "loft" as a term that sums up these urban qualities. (…) It is used to describe adaptable, flexible, and, at the same time, a powerful space. (…) The qualities of the loft are in that sense not limited to a single building—they can be transferred to the urban context as a whole". Martina Baum, Kees Christiaanse, *City as loft*. gta Verlag, ETH Zurich, 2021, p. 9/10.

There are endless opportunities for different uses, practices, experiences that follow one another in the same rethought space as well as a succession of as many *in-between* spaces (see p. 26 in Hertzberger 1991). The project creates *inescapable reality* (see p. 17 in Lefaivre and Tzonis et. At Voelker 1999) that evolves.

In fact, through its drawing, the project can build as many different *scenarios* or invite with an open use, functions through recipients that are no longer *users*, but *dwellers* (Hertzberger 1991). They are now main actors in detecting the city's potential, as "whatever space and time mean, place and occasion mean more!" (see p. 121 in Van Eyck 1966).

Each device that makes the city is the bearer of meanings only when it comes into action with other elements, becoming a part of the geometry that is no longer static but changeable/fluid.

As Hertzberger says, the devices are rendered as a mechanism that is able to detect the energy of the project. In the instant when they enter into a reciprocal relationship, they give life to another, they strip themselves of their original meaning and take on a different value.

They become gears of a sought reciprocity.

In this perspective, space becomes a structure that consists of infinite probabilities. It turns into a structure that is governed by change (Balmond 2002) of adaptation and change and that reinforces the idea that "the real dimension of the city is not space but time" (see p.7 Doxiadis 1974) and the project "the scene of action (…) is not three-dimensional Euclidean space but rather a four-dimensional world, with space and time (…), linked together (see p. 118 in Strauven 1998).

This dynamic vision of the project encourages the experiments that develop an architecture without fixed adjuvants. It calls for impermanent, flexible, incremental architectures that incorporate the concepts of adaptability, temporariness, interaction, and co-presence.

Those structures are able to incorporate the *non-plan* as well as including continuous transformation and free interaction between users in space.

Projects such as *Fun Palace* (1961) and *Potteries Thinkbelt* (1964), both by Cedric Price, are multi-purpose three-dimensional projects and structures that help reinforce the idea that architecture should be thought of as a liberating action capable of anticipating, producing and of stimulating actions.

This change is possible through modifiable, incremental, adaptable, mobile structures that establish precisely such open dialogue between space and time.

At the same time, "spatial infrastructure': a multi-storey space-frame-grid, which is supported by widely-spaced piles […]. This infrastructure forms the fixed element of the city. The mobile element consists of walls, base-surfaces and dividing walls which make the individual division of the space possible; it could be called the 'filling' for the infrastructure. All elements, which come into direct contact with the users (i.e. those they see, touch etc.) are mobile, in contrast to the infrastructure, which is used collectively and remains fixed." (Friedman 1960), such as Constant's *New Babylon* (1956) at Friedman's *Ville Spatiale* (1959), imagines now how addable layers generate a level parallel to the city space in constant evolution, without constraints, both planimetric and internal, through flexible devices.

The interior is continuously redefined in its form, in its boundaries, in its use.

The result is a space of incremental intensity where the user becomes an organizer of the space evolution according to his needs. Architecture's aim is to determine a *carrying order* (Smithson 1974), to come up with an infrastructural project that is capable of variation and growth and that follows the principle "function follows form" (Fridman 2000).

"First of all, opposite to buildings which can have a "form", cities are formless. Secondly, as for "function", cities have no precise function: they are "multi-functional". Multi-functionality can go hand in hand with formlessness would be the urban analogue to "function follow form". The role of users, as form urban reality, is all-important and totally unconscious" (see p. 112 in Fridman 2000).

Visionary *pop-up* architectures also emerge.

Structures that punctually inject into the urban fabric a series of playful and cultural functions to temporarily intensify the use, the relationships in that portion of the city.

For example, the macro flying structures *Instant City* by the Archigrams (1968) and the independent microlight structures *Magnets* by Cedric Price (1996) are based on the idea that "the role of architecture shifted from its traditional task as hardware (…) to that of designing "software" that would enable different social situations in given space" (see p. 135 in Steiner 2000).

These mobile devices, conceived as pieces of a *generative grammar* (see p. 93 in Hertzberger 1991), act on the existing space with a *random* effect.

They are conceived as extraneous elements that load the space, discovering as many available spaces and the relative degrees of intervention contained in them. They determine a new balance and a different way of using and living in the city they underline the idea that the project should be thought of as a *generator of potential*—as a catalyst that *anticipating the unexpected* (Cedric Price 1996).

Therefore, such mobile devices are able to stimulate the alternative use of space and the participation of the inhabitants, encouraging exploration, processes that intensify the use and the relationship with the urban space.

Altogether, these subversive works act as an activator to overcoming the filters, disrupting and subverting the fixed situations. Through ephemeral and mobile structures, interchangeable and incremental, the *Invisible Sandwich* of the urban context is revealed. This concept "alludes to all of the gap-like potentials that architects must be able to grasp, both in terms of time and space. (…) an invisible dictionary of possibilities that lets you redefine rather than improve or enhance a context" (see p. 13 in Price 2003).

In this view, the project and its devices are able to facilitate something new in the same place, taking the initiative, acting as the promoter to include the person who gets things moving, or in other words, it presumes or admin the capability of acting otherwise (see p. 220 in Giddens 1987).

At the same time, the project becomes a space for "*both-and*" rather than a space for "*either/or*" (Venturi 1966) because "an architectural plan is only the triggering of a long-wearing erratic dynamic process" (see p. 111 in Fridman 2000).

In this way, projects, as such, represent a significant moment in themselves, but their quality derives as much from what was built before, as from what it is able to/can become.

Thanks to simple devices, these spaces accept the unexpected and can inspire an idea or allow for building new, without interruption, the possibilities of the future.

As a process that generates a space as an undefined place.

4.5 Tools for Intensity

The intensity spaces are thus activated through their design and devices.

Analyzed as individual components of "complex assemblage", these devices reveal the relationships that they establish, and the ones that they generate.

That is how they permanently leave room for a potential change.

According to this approach, this section is structured as a practical toolkit that aim to highlighting and to name strategies and project actions that characterize the intensity spaces as an activators' space, through a different scale: area/context, site/building, object/device.

Recreating the space

Intensity spaces need to be adaptive in terms of their space and time scale thanks to the flexible devices that are able to adapt to the diversity and the evolving use, functions, and practices.

These spaces give their users the means to act on the very fabric of these spaces to provide possibilities and to find other ways of doing things.

Space in a state of permanent transformation is able to become a laboratory of inventions and experiments where new solutions can be invented and tried out in a kinetic time.

Blurring the limits

To transforms the inaccessible and unsuited spaces or too restrictive in a place.

Intensity spaces are defined but they are not limited. It is that space that includes the possibilities to let itself be crossed. In their turn, the devices are able to transform the inaccessible or too restrictive space in free space, indeed in a intense place.

The free space is defined but not enclosed thanks to the open or hybrid external boundary that changes the nature of the space and creates the feeling of possibility to act in intensity places.

Working on the outer boundary—open or hybrid—changes the nature of the space and creates or accompanies the feeling of taking part in the intensity places.

It is possible to work on the boundaries and borders as ambiguous and porous edges, as Sennett says. The boundary is an edge where things end; the border is an edge where different groups interact. On borders, organisms become more interactive, due to the meeting of different species or physical conditions—for instance, the difference between wall and membranes.

While walls retain as much as possible internally, the membranes are both porous resistant at the same time. So, we should aim to use the devices for the accessibility of working as a border/membrane.

Hybrid thresholds

Intensity space as a space of *coexistences.*

Free space can be conceived as a continuum of undefined space where different *users* or *dwellers* coexist over the same platform.

Mixed use

To think of the space as a platform within which the multiplicity of uses is included.

This dynamic configuration is produced by the devices but also through many uses by individuals connected in the same space-time.

This is what makes the space an intense and proactive space.

As Jane Jacobs says, "a mixture of uses, if it is to be sufficiently complex to catalyst public contact and cross-use needs an enormous diversity of ingredients" (see p. 144 in Jacobs 1961).

Capability of action

A project, as such, represents a significant moment in itself, but its quality comes as much from what has been built before but more as what is able yet to come.

Open spaces are activated, modified, and transformation by the people, who permeate them with new potential. In this process, the actions become experiences.

Intensity space is designed to encourage free actions and explore the possibility of activating it and recreating it, open to an unlimited number of changes. At the same time these uses relate to the availability and flexibility of things depend equally on the initiative of the individual.

This is to say that uses include a dimension of discovery in the presence of things that are limitless, incomplete, infinite. As Gilbert Simondon says in *Cours sur la perception:* "The modality of what is imagined being that of what is potential, it does not become that of unreality unless the individual is deprived of access to the condition for its realization".

Unfinished form

Intensity space is a space with infinite possibilities, limitless in time and space, as well as it is an undefined space.

Thanks to the drawing, it is a space accomodate the unexpected in order to construct, uninterrupted possibilities on the future and may spark an idea or facilitate something new. It is planned so that it can be redefined during the time and adapt to new purposes, as well as the needs of change thanks to more flexible devices, as a part of the provoke structures that invite the free act and social relationships. In fact, the realization of an endless series of interactions between the actual use and the hidden potential of space and the individual action can be found if we rethought the space as a un-prescriptive process.

Take actions

All the uses depend on certain things, however to generate an intensity spaces it is crucial to set conditions in which things are not only available but also at least to some extent, flexible, able to adapt to the individual needs. Infinite spaces and infinite uses rely on the availability and flexibility of open devices that are never complete or fixed but, on the contrary, are susceptible to variation depending on the situation in which they are called.

Following the Cullen's concept "art of the relationship" the intensity spaces are therefore based on the strengthening of the relationship that the different elements can trigger according to their design. Because "the juxtaposition effects are in themselves just as exciting as the juxtaposed objects themselves—often even more" (see p. 183 in Cullen 1961).

Take part

Intensity space is thinking, and designing a space able to extends its influence outside its boundaries. In this way, space thus becomes an environment and promotes an atmosphere that extends beyond the actual building.

This type of functionality multiplies and intensifies the interaction between people and places, and indeed it creates a space that is defined, open and able to melt into a surrounding urban behavior (Fig. 4.1).

4.6 Conclusion

Intensity, thanks to its ability to re-read the space in its space-time dimension, is able to make show up its thickness.

Thickness is given by flows, rhythms, trajectories, but at the same time by actions and practices—all processes in progress that force to rethink and describe space as a kinetic frame.

In this different perspective, all spaces become the catalysts and activators of multiple actions. In fact, "Architecture is not simply about space and form, but also event, action, and what happens in space" (Tschumi 1994).

Whether it is a room or a part of the city—in fact any kind of space at all, following certain strategies and adapting diverse devices—can evolve to be more "sticky" (Stonor 2018) being used more intensively.

These strategies and devices augment participation and therefore also enhance experiences.

Fig. 4.1 Tools for intensity

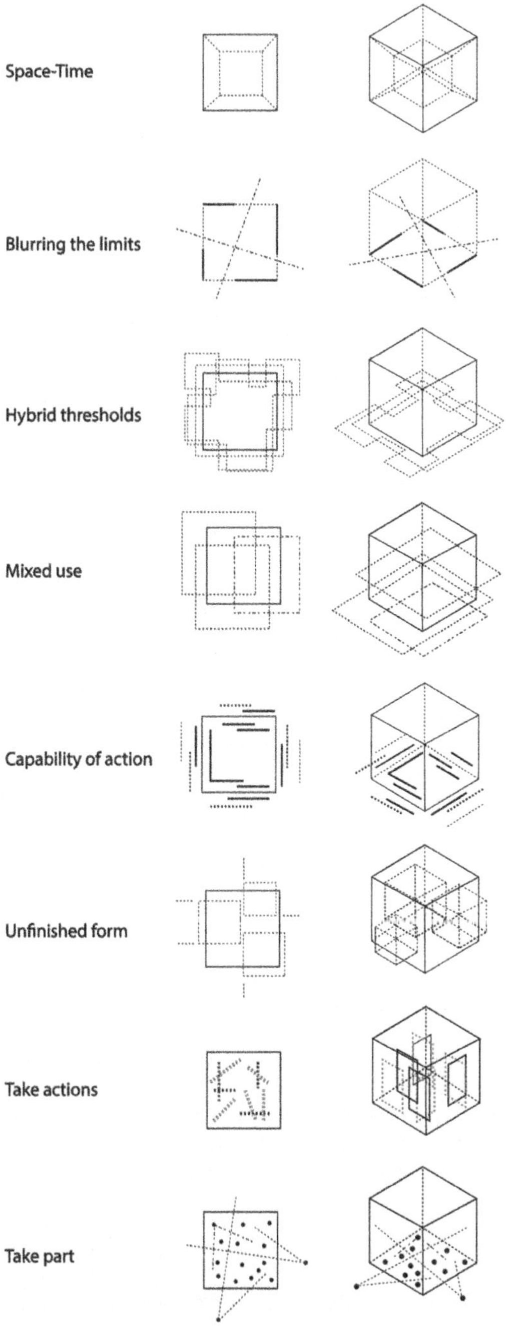

References

Alexander, C. (1965). A city is not a tree. In *Architectural forum*, 122(1), 58–62 (Part I), 122(2), 58–62 (Part II).

Amin, A. (2015). Animated spaces. *Public Culture, 27*(2), 239–258.

Amin, A., & Thrift, N. (2002). *Cities. Reimagining the urban*. Cambridge: Polity Press.

Amphoux, P. (2003). Polarité, Mixité, Intensité, Trois dimension conjointes de la densité urbaine. In H. Heyner & D. Vanderburg (Eds.), *Inside Density, International Colloquium on Architecture and Cities* (pp. 19–32). Bruxelles: La lettre volée.

Arent, H. (1958). *The human condition*. Chicago: The University Chicago.

Balmond, C. (2002). *Informal. Architecture Series*. Munich: Prestel.

Baum, M., & Christiaanse, K. (2012). *City as loft: Adaptive reuse as a resource for sustainable urban development*. Zürich: gta Verl.

Bachelard, G. (1932). *L'intuition de l'instant*. Paris: Stock.

Bauman, Z. (2000). *Liquid modernity*. Cambridge: Polity Press.

Botsman, R., & Rogers, R. (2010). *What's mine it's yours. The rise of collaborative consumption*. New York: Harper Collins.

Cedric, P. (1996). Anticipating the unexpected. *Architects' Journal*, 27–39.

Cedric, P. (2003). The invisible sandwich. In H. U. Obrist (Ed.), *Re:CP cedric price*. Basel: Birkhauser.

Cullen, G. (1961). *Townscape*. London: Architectural Press.

De Certeau, M. (1980). *L'Invention du Quotidien*. Paris: Union générale d'éditions. English edition: De Certeau, M. (1984). *The practice of everyday life* (Rendall, S., Trans.). Berkeley: University of California Press.

De Certeau, M. (1984). *The practice of everyday life* (Rendall, S., Trans.). Berkeley: University.

De Landa, M. (2005). Space: Extensive and intensive, actual and virtual. In I. Buchanan & G. Lambert (Eds.), *Deleuze and space* (pp. 80–87). Edinburgh: Edinburgh University Press.

Deleuze, G. (1968). *Différence et Répétition*, Paris: Press Universitaires de France. English edition.

Deleuze, G. (1970). *Spinoza—Philosophie pratique*. Paris: Les Éditions de Minuit. English edition.

Deleuze, G. (1988). *Spinoza: Practical philosophy* (Hurley, R., Trans.). San Francisco: City Lights Books.

Deleuze, G. (1994). *Difference and Repetition* (Patton, P., Trans.). New York: Columbia University Press.

Deleuze, G., & Guattari, F. (1980). *Mille plateaux. Capitalisme et schizophrénie*. Paris: Les Editions de Minuit. English edition.

Deleuze, G., & Guattari, F. (1987). *A thousand plateaus: capitalism and schizophrenia* (Massumi, B., Trans.). Minneapolis: University of Minnesota Press.

Dovey, K. (2012). Informal urbanism and complex adaptive assemblage. *International Development Planning Review, 34*(4), 349–368.

Doxiadis, C. (1974). *Ecumenopolis: The Inevitable City of the Future*. New York: W.W Norton & Company.

Foucault, M. (1967). Des espaces autres. *Architecture/Mouvement/ Continuité, 5*, 46–49; English edition: Foucault, M. (1986). Of other spaces (Miskowiec, J., Trans). *Heterotopias, Diacritics, 16*, 22–27.

Foucault, M. (1986). Of other spaces (Miskowiec, J., Trans). *Heterotopias, Diacritics, 16*, 22–27.

Friedman, Y. (1960). Architecture mobile. In Eaton, R. (Ed.). (2002). *Ideal cities: Utopianism and the (Un)built environment*. London: Thames & Hudson.

Fridman, Y. (2000). Function follow form. In H. Hughes & S. Sadler (Eds.), *Non-plan: Essays on freedom participation and change in modern architecture and urbanism* (pp. 104–113). Oxford: Architectural Press.

Gausa, M., Guallart, V., Muller, W., Soriano, F., Porras, F., & José, M. (2003). *The Metapolis dictionary of advanced architecture: City, technology and society in the information age*. Barcelona: Actar.

Gehl, J. (1980). *Livet mellem husene: udeaktiviteter og udemiljøer*. Copenhagen: Arkitektens Forlag; English edition: Gehl, J. (1987). *Life between buildings*. (Koch, J., Trans.). New York: Van Nostrand Reinhold.
Giddens, A. (1987). *Social theory and modern sociology*. Cambridge: Polity.
Goffman, E. (1967). *Interaction ritual*. Chicago: Aldine.
Harvey, D. (1990). *The condition of postmodernity*. London: Basil Blackwell.
Heidegger, M. (1954). *Bauen, Wohnen, Denken*. Pfullingen: Gunther Neske Verlag. English edition: Heidegger, M. (1971). *Building, Dwelling Thinking*. In *Poetry, language, thought* (Hofstadter, A., Trans.) (pp. 145–161). New York: Harper & Row.
Hertzberger, H. (1991). *Lessons for students in architecture*. Rotterdam: NAi010 Publishers.
Jacobs, J. (1961). *The death and life of great American Cities*. New York: Random House.
Lefaivre, L., & Tzonis, A. (1999). *Aldo Van Eyck: Humanist Rebel*. Rotterdam: 010 Publishers.
Lefebvre, H. (1974). *La production de l'espace*. Paris: Anthropos. English edition: Lefebvre, H. (1991). *The production of space*. (Nicholson-Smith, D., Trans.). Cambridge, MA: Blackwell.
Lepetit, B., & Pumain, D. (1993). *Temporalités urbaines*. Paris: Anthropos.
Ligtelijn, V. (1999). *Aldo Van Eyck*. Basel: Birkhauser Verlag.
Lynch, K. (1960). *The image of the city*. Cambridge: MIT Press.
Mahler, J. (2005). *Ladies and Gentlemen, the Bronx Is Burning. 1977, Baseball, Politics, and the Battle for the Soul of a City*. New York: Farrar, Straus and Giroux. *Living Big in a Loft* in LIFE, 27 March 1970, pp. 61–65.
Massey, D. (2005). *For space*. London: SAGE Publications.
McFarlane, C. (2011). Assemblage and critical urbanism. *City, 15*(2), 204–225.
McFarlane, C. (2015). The geographies of urban density: Topology, politics and the city. *Progress in Human Geography, 40*(5), 629–648.
O'Doherty, B. (1986). *Inside the white cube, the ideology of the Gallery Space*. Santa Monica: Lapis Press.
Oswalt, P., Overmeyer, K., & Misselwitz, P. (2013). *Urban catalyst: The power of temporary use*. Berlin: DOM publishers.
Salat, S., Labbé, F., Nowacki, C., & Walker, G. (2011). *Cities and forms: On sustainable urbanism*. Paris: CSTB Urban Morphology Laboratory.
Sassen, S. (2013). Informal economies and cultures in global cities. In P. Oswalt, K. Overmeyer, & P. Misselwitz (Eds.), *Urban catalyst: The power of temporary use* (pp. 104–116). Berlin: DOM publishers.
Sennett, R. (2008). *The open city*. www.richardsennett.com.
Sennett, R. (2010). The public realm. In G. Bridge & S. Watson (Eds.), *The blackwell city reader*. Wiley-Blackwell: Hoboken.
Sennett, R. (2018). *Building and dwelling: Ethics for the City*. London: Allen Lane.
Simmel, G. (1908). *Soziologie*. Leipzig: Duncker & Humblot. English edition: Simmel, G. (2009). *Sociology: inquiries into the construction of social forms*. (Blasi, A. J., Jacobs, A. K., & Kanjirathinkal, M., Trans.). Boston: Brill.
Simmel, G. (2009). *Sociology: Inquiries into the construction of social forms*. (Blasi, A. J., Jacobs, A. K., & Kanjirathinkal, M., Trans.). Boston: Brill of California Press.
Simondon, G. (1964). *L'individu et sa genèse physico-biologique*. Paris: Presses Universitaires de France.
Smithson, A. (1974). How to recognize and read mat-building. *Architectural Design, 44*(9), 573.
Smithson, A., & Smithson, P. (1953). *An urban project*. In Dannatt, T. (Eds.). *Architects' Year Book 5*. London: Elek.
Smithson, A., & Smithson, P. (2005). *The charged void: Urbanism*. New York: The Monacelli Press.
Solomon, A. (1967). *New York: The new art scene*. New York: Hort Rinehart Winston.
Sorkin, M. (2003). Density noodle. In *Lotus International*, 117, pp. 4–12.
Steiner, H. (2000). Off the map. In H. Hughes & S. Sadler (Eds.), *Non-plan: Essays on freedom participation and change in modern architecture and urbanism* (pp. 126–137). Oxford: Architectural Press.

Stonor, T. (2018). *Intense relationships: measuring urban intensity*. In *Intensity, Architectural Review*, 1451, pp. 22–25.
Strauven, F. (1998). *Aldo van Eyck. The shape of relativity*. Amsterdam: Architectura & Natura.
Thrift, N. (2007). *Non-representational theory: Space, politics, affect*. London: Routledge.
Thrift, N. (2014). The 'sentient' city and what it may portend. *Big Data & Society, 1*(1). https://doi.org/10.1177/2053951714532241.
Tonkiss, F. (2013). *Cities by design: The social life of urban form*. Cambridge: Polity Press.
Tschumi, B. (1994). *The manhattan transcripts*. MT1: Academy Editions.
Tuan, Y. (1977). *Space and place. The perspective of experience*. Minneapolis: University of Minnesota Press.
Van Eyck, A. (1966). *Labyrinthine clarity*. London: World Architecture Three.
Venturi, R. (1966). *Complexity and contradiction in architecture*. New York: The Museum of Modern Art.
Whyte, W. H. (1980). *The social life of small urban spaces*. New York: Project for Public Spaces.
William, R. (1976). *Keywords. A vocabulary of culture and society*. New York: Oxford University Press.
Zukin, S. (1982). *Loft living. Culture and capital in urban change*. Baltimore: John Hopkins University Press.

Chapter 5
The Appraisal Challenge in Cultural Urban Regeneration: An Evaluation Proposal

Beatrice Mecca and Isabella M. Lami

Abstract In the recent decades, cities need to face new challenge in environmental, economic, social and cultural dimensions and the awareness of the necessity of significant intervention able to address these conditions, led to identify urban regeneration policies to tackle these problems in all their multidimensionality. Moreover, recently in this context, culture becomes to be used as a tool for improving the conditions and the changes of many cities, in order to avoid the demolition of industrial sites, urban centres and abandoned areas and to handle their new urban trajectories. Culture became the new way to renew the image of cities or districts, but the key question is how we can measure the impacts of this phenomenon. According to this, starting from a literature review of evaluation procedures in the urban regeneration fields, the aim of the chapter is to provide a proposal of evaluation in order to assess the outputs of a regeneration process based on culture. Through a combination of approaches, interfacing with experts, the chapter seek to provides a set of indicators with the aim to assess the value-added by the regeneration process according to environmental, socio-cultural and economic fields.

5.1 Introduction

In the last 30 years, many cities need to face new challenges according to the social and economic changes, which entails new structure on social, cultural and economic dimensions, from a material and immaterial point of view (Ferilli et al. 2016). At the broadest level, the awareness that cities need significant intervention in order to improve their conditions, arises from a particular understanding of the causes of the physical, economic and social problems of the cities themselves.

B. Mecca (✉) · I. M. Lami
Department of Regional and Urban Studies and Planning (DIST), Politecnico di Torino, Viale Mattioli 39, 10122 Turin, Italy
e-mail: beatrice.mecca@polito.it

I. M. Lami
e-mail: isabella.lami@polito.it

© Springer Nature Switzerland AG 2020
I. M. Lami (ed.), *Abandoned Buildings in Contemporary Cities: Smart Conditions for Actions*, Smart Innovation, Systems and Technologies 168,
https://doi.org/10.1007/978-3-030-35550-0_5

In this context, urban regeneration has to define the appropriate policy to tackle all these problems in all their multidimensionality (Abastante and Lami 2013), considering the interactions among the problems and the nexuses that led to a decline of the cities (De Magalhaes 2015). Urban renewal is based on the assumption that social and economic problems are not a-spatial and, also according to the concept of the adaptive reuse, the character of location determines the nature of these problems (Lami et al. 2014). So, the multidimensionality of the urban regeneration's themes entails objective on different fields: economic, social, cultural and environmental (Tavella and Lami 2019; Lami and Tavella 2019).

In the last decades "culture" represents a driving factor of the urban regeneration processes. It becomes the triggering element that allow to address the complex and huge problems of industrial sites, urban centres and abandoned areas, by renewing the image of the city and of its districts, encouraging the sense of belonging of residents, improving the social cohesion, giving benefit to health and environment, attracting new investments and tourist flows and producing new jobs in the cultural sectors (Ferilli et al. 2016).

Since we recognise that in almost all the six cases of urban regeneration illustrated in the book, culture is considered as the key element of the entire process (as in several European interventions), the main challenge of this chapter is to understand how we can assess the achievements of the goals, the benefits of the renewals in the economic, socio-cultural and environmental fields. We start from an analysis of the main evaluative procedures performed in the context of urban regeneration, then we move to a literature review of the tools, techniques and methods used in the field of cultural urban regeneration.

According to this, Sect. 2 analyses the urban regeneration process as a wicked problem and the related evaluation procedure, then Sect. 3 provides a description of the cultural urban regeneration phenomenon, followed by Sect. 4 that analyses the evaluation procedures currently suggested in literature on cultural urban regeneration. Section 5 provides the proposal of an evaluation framework based on a set of environmental, socio-cultural and economic indicators for defining the possible added-value of the area. Finally, Sect. 6 summarizes the conclusions.

5.2 Evaluation Method for Urban Regeneration

The urban regeneration phenomenon can be explained as a comprehensive and integrated action that aids to solve urban problems, seeking to led to a lasting improvement in the economic, social and environmental condition of an area subject to change (Roberts 2000). This phenomenon arises from the transformation of the urban historical or industrial areas in order to avoid the perception of the decline of the city and to allow the possibility of recovery and revitalisation (Amistà and Cerreta 2018; Abastante et al. 2020). Considering the unused or underused buildings and areas as a resource to be activated, we can approach these goods adjusting their form and functions in a long-term (Salat 2011) or adopting the adaptive reuse (Németh and Langhorst 2014).

Urban renewal is a multidimensional issue that entails objective on different fields: economic, social, cultural and environmental. According to this, it can be identified as "wicked problem situation" (Rittel and Webber 1973), namely a complex problem without a simple method of solution; they are considered problems for which there is no definite end point and on which we cannot affirm with certainty that it has been solved. As Wang et al. (2012) stated, different studies demonstrate that local values and identities and urban regenerations have become significant issues. These include physical requalification, economic growth, encouraging social inclusion and reducing inequalities and increasing community cohesion without forgetting the environmental issue, that have great importance following climate change (De Magalhaes 2015). Recently, transforming urban historic areas has become increasingly complex because of a number of reasons, including the property fragmentation, the values' creation mechanisms, the regulatory framework, the multitude of public and private stakeholders with divergent perspectives and values (Lami 2019). The problem situations related to urban regeneration can be recognised as a "super wicked problem", borrowing the term from the literature on climate change (Lazarus 2009) and introducing it urban planning and evaluation realm. "It is possible to distinguish a series of issues that can grouped in three main features: (i) the measurable dimension of the problem, mainly related to the physical and spatial dimension of that peculiar economic good represented by the urban tissue; (ii) the specificity of the decisional processes in this realm; (iii) the normative dimension" (see p. 251 in Lami 2019).

Since the aforementioned nature of urban regenerations, specific evaluation procedures are needed, to consider the complexity analysing different factors, from the economic and spatial ones to the functional and environmental aspects. Lots of tools and techniques are available for the assessment of the process respect to the three phases of the project, namely: the *ex ante* phase, in which the evaluation techniques are used before the preparation of the project in order to support the formulation of the project; the *in-itinere* phase, in which are used techniques related to the control of the initial objectives, highlighting the effects during the construction phase; finally the *ex post* phase that entails tools for learning from the experiences and for informing the actors involved of the results attained. Table 5.1 provides an overview of four types of evaluation analysis used in the context of urban regeneration, the Discounted Cash Flows—DCF (Hoesli 2000), the Cost-Benefits Analysis—CBA (European Commission 2014), the Social Return of the Investment—SROI (Millar and Hall 2013) and the Multicriteria Decision Analysis—MCDA (Figueira et al. 2005).

Most of the applications related to urban renewals concern the use of MCDA, or a combination of economic evaluations and multicriteria ones. See for example: a nominal classification method to define the adaptive reuse of abandoned buildings for cultural uses (Costa et al. 2019a) or to assign a house to refugees (Costa et al. 2019b); the evaluation of social housing initiatives putting together a parsimonious AHP methodology and the Choquet integral (Abastante et al 2018); an integrated assessment framework (combining stakeholders analysis, Strategic Choice Approach

Table 5.1 Overview of four evaluation tools for measuring the impacts/benefits of urban regeneration

	DCF	CBA	SROI	MCDA
Description	The DCF is an economic and financial tool for the assessment of the income of an investment in monetary terms. Respect to a defined cash-flow period it provides two profitability indicators of the investment	The CBA is used for the analysis of the social changes due to the intervention. It tries to quantify all the costs and the benefits through monetary value, including the aspect for which the market does not provide any economic value	The SROI is a methodological tool for the assessment of the social impacts and for mapping the change. It provides advice in order to improve the quality of the urban project. It measures the benefits of the project through monetary value	The MCDA is used for the comparison of scenario alternatives and it allows to investigate the full range of aspect of a project through qualitative and quantitative information in a single assessment
Phase application	Ex ante	Ex ante	Ex ante, as a forecast of the values, or Ex post as evaluation of the results	Ex ante, as a forecast of the values, or Ex post as evaluation of the results
Strengths	This evaluation provides results easy to communicate	This evaluation provides results easy to communicate	This tool considers the social, economic and environmental benefits, so it provides more information than just the financial projections	This evaluation represents the decision-making process and according to the techniques used, it could have a good communicability
Weaknesses	The method does not take into account any externalities of the project	The assignment of the monetary value to the externalities of the project could be complex and inaccurate or unacceptable	The long and complex process for defining the monetary value of the intangible aspects could led to inaccurate values	These methods have a high level of subjectivity, indeed the results are subject to variation over time

and MACBETH) for the requalification of districts facing urban and social decline in Brussels (Abastante and Lami 2018); the application of a Spatial Decision Support System (SDSS), integrated with financial and economic evaluations, to analyse energy scenarios at district level for the development of public policies and private decisions in the real estate sector (Abastante et al. 2017); the use of Dominance-based rough set approach and analytic network process for assessing urban transformation scenarios (Abastante et al. 2013); the use of dynamic maps to support planning processes (Pensa et al. 2013).

5.3 Cultural Urban Regeneration

In recent decades, the role of culture became a driving factor in the process of urban regeneration, used as a "leverage" for improve the conditions of abandoned or underused places, in order to avoid the demolition of industrial sites, urban centres and abandoned areas and to handle their new urban trajectories. Culture assumes a decisive role for a comprehensive development of the city, improving the interaction between economic development, the social renewal and the regeneration building (Sepe 2013).

Evans (2005) defines three type of role of culture within the regeneration project:

- Culture-led Regeneration: culture is seen as an aesthetical tool, cultural project and events are used to achieve greater community participation;
- Cultural Regeneration: culture is integrated in the regeneration strategy along with other environmental, social and economic activities;
- Culture and Regeneration: in this case culture is not integrated in the plan, but it is integrated later when building have already been designed. It is a technical policy tool.

Culture becomes part of the new orthodoxy through which cities try to improve their position (Miles and Paddison 2005). It makes the difference by renewing the image of the city and of its districts, encouraging the sense of belonging of residents, improving the social cohesion, giving benefit to health and environment, attracting new investments and tourist flows and producing new jobs in the cultural sectors (Ferilli et al. 2016).

A crucial question is how to understand the nature of the impact of this phenomenon, without considering the culture as a sort of panacea for urban issues: obviously culture means something, but it cannot mean everything (Miles and Paddison 2005; Blessi et al. 2012). Culture can be translated in the inputs of the process which provides social and economic outputs and is worth noticing that the human capital should be considered as the key of the successful regeneration (Miles and Paddison 2005).

It is worth highlighting the challenge represented by the desire to assess an elusive theme such as the cultural one in an operation of reuse and enhancement of an abandoned space. An evaluation model has to reason on the versatility of the places and suitability for specific uses, as well as to build an appropriate vision that links the existing infrastructure and its potential to local and global trends (Costa et al. 2019a).

According to this, how is it possible to assess the impact of culture-led regeneration in order to understand its pros and cons more in details?

5.4 Evaluation Method for Cultural Urban Regeneration

Since we recognise in almost all the six cases addressed in the research that culture is considered the triggering element of the entire process of urban regeneration, as in several European contexts, we decided to investigate which methods are used to evaluate this specific phenomenon. A literature review has been performed, structured into keywords related to the field of cultural urban regeneration.

First, the review has been carried out through the use of Scopus and ResearchGate database. Second, the scientific journals of the last twenty years have been analysed, from 2000 to 2019. Third, specific fixed keywords, such as "urban regeneration" and "evaluation/appraisal/measurement" have been used in order to restrict the research to the urban regeneration appraisal fields. Moreover, five additional keywords have been selected in order to identify the aspect we wanted to analyse, namely: indicator, human/social/cultural capital, value, benefit, impact and culture-led regeneration.

The first step of the analysis has led to 68 scientific papers of the past twenty years, from 2000 to 2019. Figure 5.1 shows the number of the papers according to the keywords identified.

From Fig. 5.1 emerges that, among the papers identified, the majority is related to the keyword "value" (19 papers) and "human/social/cultural capital" (14 papers), followed by the keywords "Impact", "Benefit", "Indicators" and "Culture-led regeneration" (with respectively 12, 10, 10 and 3 papers).

In the second step of the review, scientific papers were analysed in order to identify the ones that provides examples of evaluation approaches of the added value

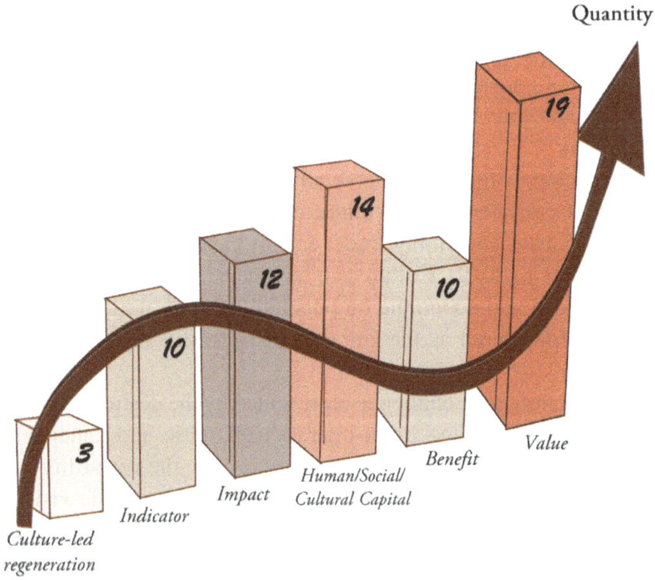

Fig. 5.1 Literature review according to the keywords identified

Table 5.2 Evaluation procedures in relation to their context application

Evaluation procedure	Context application	New use	References
Indicator-based approach	Metropolitan district/area	Circus pole, cultural centre, university, residential area	Ferilli et al. (2016), Blessi et al. (2012)
SIPRIUS (indicator system)	Industrial area	Housing, offices and educational buildings	Laprise et al. (2015)
Holistic collection of indicators (IOT, VCA, MFA)	Theoretical example of evaluation	–	Carius et al. (2018)
Public data-based approach	Urban areas	Residential and social housing area	Mak and Stouten (2014)
Cost-benefit evaluation approach	Different case in England	Improvement of industrial, commercial infrastructures	Tyler et al. (2012)
Multidimensional and multi-dimensional decision-making process	Historic district	Three scenario alternatives	Amistà and Cerreta (2018)

directly connected to culture in the urban regeneration field. Table 5.2 shows the main evaluation procedures analyzed respect to their context application and to the new uses obtained after the regeneration process.

As it arises from the Table 5.2, the regenerations tackle by the selected papers refer to areas of intervention which are sufficiently extensive (industrial area, district, metropolitan area) and the new uses refer to housing area, cultural poles and educational activities. We can also distinguish four macro-families according to the methods or techniques performed.

First, the analysis carried out with performance indicators (Ferilli et al. 2016) provides a comprehension and a qualitative description of the regeneration process through the perception of the communities, but it does not follow any standard or reference for cultural regeneration studies on social and human capital, so it addressed for a "pick up mix" approach. Accordingly, this approach will be always related to a subjective concept, which can be seen as its main weakness. Also the SIPRIUS approach (Laprise et al. 2015) uses indicators and it raises a stronger awareness on the potential of the disused urban area or buildings in the fields of urban regeneration. The transparency and the readability of the results mean to avoid a global rating which derives from quantitative aggregation of heterogeneous data such as the energy consumption and the flexibility or the security of the buildings. It is worth noticing that there is not a recipe which can be used for any project: the method is based on objectives to aim, and not on predefined means for obtain acceptable level of sustainability. The regional value added RVA method (Carius et al. 2018) represents an appropriate and concrete tool for holistic sustainability evaluation for supporting decision-making process, since it enables the comparison of alternative of project considering their regional impacts. However, the use of indicators can be both a blessing and a course: they are easy to understand and communicate but they could be not usable for every local context and can mislead in case of incomplete information.

Second, the ex-post evaluation of Mak and Stouten (2014) is interesting and apparently simple technique in its applicability, since it is based on real data provided by the municipality, however this strength could be also its greatest weakness. The availability and presence of municipal indicators is not so obvious in all contexts and countries, to the point that it may not be applicable.

Third, Tyler et al. (2012) provides monetary quantifications of impacts for the evaluation of the benefits of urban regeneration policies. As we highlighted in Sect. 2, the monetary techniques (e.g. revealed and expressed preferences method and hedonic price method) are widely used in project appraisal, since they provide evaluations and results that are easy to be communicated. However, the monetization of the elements could be reductive and sometimes inacceptable.

Fourth, the evaluative approach structured through a multi-dimensional and multi-stakeholder decision-making process proposed by Amistà and Cerreta (2018) allows to investigate the full range of aspect of a project through qualitative and quantitative information in a single assessment. This approach is able to represent the decision making and according to the techniques used, it could have a good communicability, even if its high level of subjectivity can be seen as weakness.

5.5 A Proposal of an Evaluation Procedure

The regeneration based on culture would require the use of methodologies, that combine multiple approaches to assess social, economic and environmental impacts, however these have not yet been completed developed (Evans 2005), as also confirmed by the above literature review.

Looking at the evaluative tools for the urban regeneration and at the ones for the cultural urban regeneration synthetically described, we can highlight that the latter are much more near to the techniques used in the *in-itinere* phase, such as the performance indicators, the document analysis and the monitoring data. According to the fact that the regeneration processes of the six-case study of this research can be considered ongoing processes, due to several factors, we decided to seek to provide a set of evaluative performance indicators highlighting the effects of the process.

First of all, we focused on the concept of "*intensity*", widely explained in the chapter of Baima and Robiglio. Considering the generic definition of this word, namely the power, the energy and the incidence of a phenomenon, we seek to find a way for describing the reaction of the population to the urban transformation. The aim is to show how much the dynamism of the forces of the cultural regeneration process influenced the citizens. According to this and following the indicator-based approach of Blessi et al. (2012) and of Ferilli et al. (2016), we propose to investigate the personal perception and participation of the residents and users of the spaces subject of transformation, investigated through questionnaires and/or interviews.

The comparison between the situation before and after transformation will highlight the improvement of the service according to the personal opinion of the people involved; it will also convey the idea of the intensity with which the phenomenon has been received by the population in terms of percentage of value-added.

In a second step, we propose to combine these seven socio-cultural indicators, that allowed to assess both the intangible impacts of the regeneration process, with other indicators of a much more tangible nature, based on SIPRIUS indicator system (Laprise et al. 2015). Starting from the detection of the criteria most significant and pertinent indicators to the context of the cultural regenerations observed in the literature review, we defined a set of indicators concerning the environmental, socio-cultural and economic fields. In particular, with the help of experts in the field, we identified two macro-criteria, namely the context and project one, to which belong respectively three criteria: the environmental, socio-cultural and economic ones. For each criterion, sets of indicators have been identified, for a total of 31 indicators, that will allow us to evaluate the phenomenon. As shown in Table 5.3, respect to the context criteria we defined five environmental indicators, nine socio-cultural indicators and three economic ones; while according to the project criteria we identified three environmental indicators, eight socio-cultural and three economic ones.

Considering the definition and the measurement of the tangible indicators, they can be evaluated through calculation, measurement on maps or city plans, qualitative analysis or specific methodologies based on regulations and project guidelines. Instead the intangible aspects, such as the perception and the participation, can be investigated through questionnaires and/or interviews: people will be asked to express their opinion respect to a qualitative scale.

For each indicator, both of context criteria and project ones, is provided a graphical representation, such as a histogram, that lets to communicate efficiently the difference between the initial situation and the current one or the expected situation and the current one. Furthermore, we can observe if the regeneration process creates an added value, intended as improvement of the initial situation or of the expected situation. So, it is possible to highlight the number of indicators that gets a value greater than the thresholds and to define a percentage of value-added for each criterion. In the end these percentages can summarize in order to provide a total percentage of value-added respect to each macro-criterion (environmental, socio-cultural and economic).

5.5.1 Context Criteria

Starting from the context criteria it is provided an overview of the description of all the indicators considered, with the information relative to the evaluation method and the measurement units.

Table 5.3 Total set of indicators

Macro-criteria	Criteria	Indicators	
Context	Environmental	C_E.1	Quality of service in public transport
		C_E.2	Number of parking spaces
		C_E.3	Junction density of cycle mobility network
		C_E.4	Average emission of noise e day
		C_E.5	Average emission of noise e night
	Socio-Cultural	C_S-C.1	Number of art school
		C_S-C.2	Number of coworking spaces
		C_S-C.3	Number of public parks
		C_S-C.4	Number of recreational green/natural area
		C_S-C.5	Number of cultural centres
		C_S-C.6	Number of sports centres
		C_S-C.7	Community activity perception
		C_S-C.8	Community activity participation
		C_S-C.9	Places for social interaction
	Economic	C_EC.1	Net employment density
		C_EC.2	Proportion of work carried out by local companies
		C_EC.3	Local employment opportunities
Project	Environmental	P_E.1	Land use
		P_E.2	Reuse of existing structures
		P_E.3	Materials from renewable or recycled sources
	Socio-Cultural	P_S-C.1	Green surfaces
		P_S-C.2	Quality of outdoor spaces
		P_S-C.3	Degree of security
		P_S-C.4	Degree of functional mix
		P_S-C.5	Cultural events perception
		P_S-C.6	Cultural events participation
		P_S-C.7	Educational activity
		P_S-C.8	Quality of the environment
	Economic	P_EC.1	Investment costs
		P_EC.2	Gross rental yield
		P_EC.3	Degree of flexibility of building

ENVIRONMENTAL

C_E.1—Quality service in public transport

This indicator captures the level of quality service offered by public transport and it measures the average distance to the considered stop (Laprise et al. 2015, Bioedilizia e ITC-CNR 2012).

First, the number of types of public transport (bus, tram, subway and train) should be analysed and defined. Second, for every type it will be measured the average distance to the considered stop from the site/area. Third, through the comparison between the initial situation and the one after the regeneration process, the improvement of the service can be observed and expressed in percentage.

Unit of measure: meters [m].

C_E.2—Number of parking spaces

This indicator observes the number of parking spaces planned to meet the needs of residents, workers and visitors (Laprise et al. 2015). It entails the calculation of the number of parking spaces needed according to the urban private and/or public standard.

First, the number of parking spaces should be analysed and defined according to the urban standards and then through the comparison between the initial situation and the one after the regeneration process, the improvement of the service can be observed and expressed in percentage.

Unit of measure: meters [m].

C_E.3—Junction density of cycle mobility network

Since experts and studies consider the junction density a determinant factor in the assessment of the quality of cycle mobility network, this indicator notices the level of safety and continuity of the cycle path. Defining the "nodes" as the intersection between a cycle path and a road of equal or higher hierarchical level, the junction density is calculated as the ration of the number of nodes to the length of a cycle route (Athanasopoulos and Milakis 2014). Subsequently, comparing the initial situation and the one after the regeneration process, the improvement of the quality of the junction can be observed and expressed in percentage.

$$\text{Unit of measure: } \frac{n.nodes}{cycle\,route\,length\ (\text{km})}[\text{km}^{-1}]$$

C_E.4—Average emission of noise - day

The noise exposure constitutes a major problem in urban environments, since it affects the human health and behaviour and it can chase animals out of their habitat (Maisonneuve et al. 2009). According to this, the indicator observes average noise exposure of building located within the perimeters of the project during the day: morning (08:00–12:00), afternoon (14:00–18:00).

Passchier-Vermeer and Passchier (2000) report that most of urban population in industrialized countries are exposed to >50 dB outdoor and the rural population are usually exposed to <50 dB outdoor. According to this we propose to consider an added-value respect to this indicator for the regeneration process if it does not exceed a value of 50 dB.

Unit of measure: decibel [dB].

C_E.5—Average emission of noise night

As the same as for the indicator E.4, the E.5 corresponds to the average noise exposure of building located within the perimeters of the project during the night: evening (19:00–23:00), night (24:00–05:00).

Since the noise in evening-night time is more annoying than in the day-time, according to Passchier-Vermeer and Passchier (2000) the sound levels during the night could be about 10 dB and during the evening about 5 dB. In line with this and considering the threshold defined in the previous indicator C_E.4, or 50 dB as a noise threshold preferably not to be exceeded, we propose to consider an added-value respect to this indicator for the regeneration process if it does not exceed a value of 40 dB.

Unit of measure: decibel [dB].

SOCIO - CULTURAL

C_SC.1—Number of art school
C_SC.2—Number of coworking spaces
C_SC.3—Number of public parks
C_SC.4—Number of green/natural area
C_SC.5—Numberofcultural centres
C_SC.6—Number of sports centres

These indicators measure the number of art schools, coworking spaces and cultural centre, recreational green area, public parks, and sport centre, which are the main activities proposed by the case studies analysed in the field of regeneration based on culture. These indicators allow to observe if the new regeneration builds a kind of value for the community, in terms of decreasing times and costs to reach the points of interest through the creation of new similar services.

In general, as McCormack et al. (2008) and Yong Yand and Diez-Roux (2012) report, the walkable distance has a range between 400 and 1500 m, which means a distance between 10/15 min from homes. The distance of 400 m is considered the longest walkable distance to reach a public transport station, while a range of 1500 m is better associated with a distance linked to recreational destination or to reach schools and shops.

Accordingly, since the activities considered are part of the recreational or school/work, destinations, the analysis should be developed considering a buffer size of 1500 m, or 15 min walk. Consequently, through the comparison between the initial situation and the one after the regeneration process, the improvement of the number of services within the buffer of 1500 m can be observed and expressed in percentage.

Unit of measure: [number]

C_SC.7—Community activity perception

C_SC.8—Community activity participation

C_SC.9— Places for social interaction

These three indicators try to assess more intangible aspects, such as (Blessi et al. 2012; Ferilli et al. 2016):

- The personal perception of community activities
- The personal participation in community activities
- The personal perception of opportunities for social interaction, which let to meet other people or other communities.

These data will be investigated through questionnaires and/or interviews to the residents and users of the spaces subject of transformation. They will be asked to express their opinion according to the participation in community activities and according to the perception of supply of activities for the community and of spaces for social interaction. The first will be evaluated on a qualitative scale (no participation, occasional participation, systematic participation, total participation); the latter on a qualitative scale (poor, sufficient, good, great). In the end, the comparison between the situation before and after regeneration will highlights the improvement of the service according to the personal opinion of the people involved. This possible difference of increase (or decrease) will be expressed as percentage of value-added.

Unit of measure: percentage [%].

ECONOMIC

C_EC.1—Net employment density

This indicator shows the number of jobs per hectare of buildable land (excluding natural area, such as forest, green areas, water bodies etc.) (Laprise et al. 2015).

Through the comparison between the initial situation and the one after the regeneration process, the increase (or decrease) of employment density can be observed and expressed in percentage [job/ha]:

$$\frac{\text{number of local unit employees}}{\text{surface}}$$

Unit of measure: percentage [%].

C_EC.2—Proportion of work carried out by local companies

The indicator analyses the number of the work, needed for the transformation project, which are performed by local companies. The calculation is given by the percentage of costs of works made by companies located in the concerned area respect to the total construction costs (Laprise et al. 2015). We assumed that to constitute an added-value for the community in which the area is located, at least 50% of the cost must be covered by local companies. According to this, all the quantity that exceeds 50%, considered as a reference of expected situation, is considered an added-value.

Unit of measure: percentage [%].

C_EC.3—Local employment opportunities

As the same for the indicators C_S-C_7/8/9, this indicator tries to assess another intangible aspect, namely the personal perception of opportunities for employment, through questionnaires and/or interviews to the residents and users of the spaces subject of transformation. They will be asked to express their opinion according to the perception of jobs supply (on a qualitative scale: poor, sufficient, good, great). The comparison between the initial situation and the one after the regeneration process will highlights the improvement of jobs according to the personal opinion of the people involved. This possible difference of increase (or decrease) will be expressed as percentage of value-added.

Unit of measure: percentage [%].

5.5.2 Project Criteria

Below the description of all the project criteria.

ENVIRONMENTAL

P_E.1—Land use

According to the will of encouraging the reuse of urbanized area in order to limit the land consumption and to avoid the waste of energy and materials caused by the demolitions and new constructions, this indicator evaluate the quantity (square meters) of urban surface that had been reused (Bioedilizia e ITC-CNR 2012). The calculation is given by the ratio between the reused covered area and the total urban surface of the central area or metropolitan area.

Through the comparison between the initial situation and the one after the regeneration process, the limit of the land consumption can be observed and expressed in percentage:

$$\frac{reused\, covered\, area}{\text{total urban surface of the area}}$$

Unit of measure: percentage [%].

P_E.2—Reuse of existing structures

Also in this case, pursuing the goal of encouraging the reuse of most existing buildings, discourage the demolition in presence of structures still performing, the indicators investigate the percentage of surface of the existing building that is reused (Bioedilizia e ITC-CNR 2012). We assumed that to constitute an added-value according to the need of avoiding the waste of energy and materials, at least 50% of the building must be reused and not demolished. According to this, all the quantity that exceeds 50%, considered as a reference of expected situation, is considered an added-value:

$$\frac{horizontal \frac{and}{or} inclined\, surface\, reused}{\text{total}\, horizontal \frac{and}{or} inclined\, surface}$$

Unit of measure: percentage [%].

P_E.3—Materials from renewable or recycled sources

Respect to the fact of reducing consumption of non-renewable raw materials, the indicator assesses the percentage ratio between the weight of materials from renewable sources that were used in the intervention and the total weight of the materials used (Bioedilizia e ITC-CNR 2012). We assumed that to constitute an added-value according to the reduction of the use of non-renewable materials, at least 50% of the building must be realized with materials from renewable sources. According to this, all the quantity that exceeds 50%, considered as a reference of expected situation, is considered an added-value:

$$\frac{weight\, of\, material\, from\, renewable\, sources}{\text{total weight of the material used}}$$

Unit of measure: percentage [%].

SOCIO - CULTURAL

P_SC.1—Green surfaces

This indicator observes the quantity of green surfaces that the new project provides. The calculation is given by the ratio between the designed green surfaces and the total surface area (Laprise et al. 2015).

Through the comparison between the initial situation and the one after the regeneration process, the improvement of the green surfaces can be observed and expressed in percentage.

Unit of measure: percentage [%].

P_SC.2—Quality of outdoor spaces

The quality of outdoor spaces describes features of the outdoor spaces and the nature of the external devices (seating, vegetation, play area, etc.) that contribute to the user's well-being (Laprise et al. 2015). This indicator will be measured on the basis of a qualitative scale from 0 to 5, as shown in Table 5.4.

The analysis should be done before and after the regeneration process and through the comparison between the two situations the improvement of the quality of outdoor spaces can be observed and expressed in percentage.

Unit of measure: qualitative scale.

P_SC.3—Degree of security

The degree of security defines the level of devices contributing to the neighbourhood safety, such as light in the common areas, access control, specialized staff, etc. (Laprise et al. 2015; Santangelo et al. 2013). It will be measured on the basis of a qualitative scale from 1 to 5, as presented in Table 5.5.

Pursuing the goal of creating spaces that can be defended at the urban level and considering the lighting system as a basic element and the self-help system from communities a factor that can not be quantify with tangible elements, the level 3 has been defined as the level of expected situation. Projects that present a higher level of security, this will be considered as added-value.

Unit of measure: qualitative scale.

P_SC.4—Degree of functional mix

Table 5.4 Description of the qualitative scale

Level	Description
0	The bad condition of the outdoor spaces makes it not usable
1	The outdoor spaces are not accessible to the public
2	A small number of outdoor spaces are accessible to the public
3	A large number of outdoor spaces are accessible to the public, but with few external devices for the user's well-being
4	All the outdoor spaces are accessible to the public, with some amenities for the user's well-being
5	All the outdoor spaces are accessible to the public, with a multiple facility for the user's well-being

Table 5.5 Description of the qualitative scale

Level	Description
1	The security measures are limited to a public lighting system
2	The security measures include a self-help system from the surrounding community
3	The area/building provides access system control through walls/fences and video surveillance system
4	The area/building is manned by specialized staff
5	The area/building provides controlled access through video surveillance and specialized staff

The degree of functional mix investigates the level of differentiation of project functions. For the quantification, this indicator is based on qualitative scale from 1 to 5 (Laprise et al. 2015), as shown in Table 5.6.

Comparing the initial situation and the one after the regeneration process, the difference and the improvement of the use of the building with more functions can be observed and expressed in percentage.

Unit of measure: qualitative scale.

P_SC.5—Cultural events perception

P_SC.6—Cultural events participation

P_SC.7—Educational activity

P_SC.8—Quality of the environment

As the context indicators SC.7/8/9, these four indicators try to assess intangible aspects according to the project field (Blessi et al. 2012; Ferilli et al. 2016), such as:

- The personal perception of cultural events
- The personal participation in cultural events
- The personal perception of opportunities in the educational fields
- The personal perception of the quality of the environment of the area.

These data will be investigated through questionnaires and/or interviews to the residents and users of the spaces object of transformation. First they will be asked

Table 5.6 Description of the qualitative scale

Level	Description
1	The project presents one function
2	The project has one function, with a small share of other secondary functions
3	The project has main function, with significant share of other secondary functions
4	the project has several functions that coexist randomly on the site, without being part of a specific concept
5	The project has several complementary functions that contribute to concept of functional diversity

to express their opinion according to their participation in cultural events (on a qualitative scale: no participation, occasional participation, systematic participation, total participation), then they will be interrogated about the perception respect to the supply of cultural events, of educational activities and respect to the quality of the environment (on a qualitative scale: poor, sufficient, good, great). Consequently, the comparison between the situation before and after regeneration will highlights the improvement of the service according to the personal opinion of the people involved. This possible difference of increase (or decrease) will be expressed as percentage of value-added.

Unit of measure: percentage [%].

ECONOMIC

P_EC.1—Investment costs

This indicator provides the total investment, including all project costs (land purchase, construction work, secondary costs) (Laprise et al. 2015).

To assess whether the new project generates an added value respect to this indicator, we propose to consider the average cost of an investment, estimated on the basis of market prices as the reference threshold, and to compare it with the investment cost produced by the new project. We will register an added value only if the investment cost is lower than the average market cost.

Unit of measure: [€/m^2].

P_EC.2—Gross rental yield

This indicator assesses the economic return of the investment produced by the cultural activities of the project. The calculation is based on the ratio of revenue to the total investment costs (Laprise et al. 2015). This return will be compared to the hypothetical value obtained from the ratio between the average value of rental income according to the market and the total investment cost, taken as a reference of expected situation. Through the comparison of these two situations, if the regeneration process based on culture will provide a higher economic income, this extra percentage value will be considered as added-value.

Unit of measure: [€/m^2].

P_EC.3—Degree of flexibility of buildings

The indicator considers the concept of flexibility as the set of mutable relations between events, contexts and uses of the spaces (Raviz et al. 2015). Forty (2004) defines the most common flexibility as the one that entails the movement of the elements of the building or the one provided by the use of removable elements (walls, floors, ceilings) and the advantage of the free plant.

Table 5.7 Description of the qualitative scale

Level	Description
1	The space is particularly rigid in architectural design schemes (only through demolitions and reconstructions it will be possible to define a new flexible scheme but erasing the existing system)
2	The space is made up of several rooms with a large and/or equivalent surface area that cannot be combined, but which leaves the freedom to choose the different uses
3	The space is characterized by different units that can be easily combined. They can be aggregated or left separate according to the usage or need
4	The space is completely free of bearing elements inside, allowing maximum design freedom

The heart of the flexibility is the "use" and it is relevant to investigates some aspects of the building: can it host multiple uses? Can it be adapted to users? Can it have spatial addictions? (Schneider and Till 2007). According to this, it is proposed to evaluate the flexibility through a qualitative scale from 1 to 4, as presented in Table 5.7:

The comparison between the situation before and after regeneration will highlights if the level of pre-existing flexibility had been respected or improved, or if the rigidity of the building had been made more flexible through significant architectural interventions.

Unit of measure: qualitative scale.

5.6 Conclusion

The literature review of the last twenty years confirms, according to what stated Evans (2005), the presence of few methodologies that combine multiple approaches to assess the social, economic and physical impact of urban regeneration and evince little literature supporting way of value creation assessment in relation with the urban regeneration. Pursuing the objective of proposing a potential evaluative tool for urban regeneration process, where culture plays a major role, we decided to not consider the monetary techniques due to their reductionist approach respect to the complexity of the multidimensional issue, neither the analysis through indicator based on public data due to the lack of these indicators in the Italian context, as well as other in Europe. Moreover, also the multi-dimensional decision-making process can not be use in our context, since our aim is not the choice between different project scenarios, but the assessment of the value-added by the regeneration processes. Therefore, we provide a set of socio-cultural indicators both for the context macro-criterion and for the project one, in order to assess the more intangible aspects and to convey the idea of the intensity with which the phenomenon has affected the population. Then combining these indicators with others that investigate the tangible environmental, socio-cultural and economic aspects, we propose a set of 31 indicators for assess the value-added of the regeneration process. In this sense, the systematic integration of qualitative and quantitative indicators could be suitable to identify and tackle the physical and social impact of the cultural urban regenerations.

References

Abastante, F., & Lami, I. M. (2013). An analytical model to evaluate a large scale urban design competition. *Geoingegneria Ambientale e Mineraria, 139*(2), 27–36.

Abastante, F., & Lami, I. M. (2018). An integrated assessment framework for the requalification of districts facing urban and social decline. In G. Mondini, E. Fattinnanzi, A. Oppio, M. Bottero, & S. Stanghellini (Eds.), *Integrated evaluation for the management of contemporary cities, Green Energy and Technology*. New York: Springer.

Abastante, F., Bottero, M., Greco, S., & Lami, I. M. (2013). Dominance-based rough set approach and analytic network process for assessing urban transformation scenarios. *International Journal of Multicriteria Decision Making, 3*(2–3), 212–235.

Abastante, F., Corrente, S., Greco, S., Ishizaka, A., & Lami, I. M. (2018). Choice architecture for architecture choices: Evaluating social housing initiatives putting together a parsimonious AHP methodology and the Choquet integral. *Land Use Policy, 78,* 748–762.

Abastante, F., Lami, I. M., & Lombardi, P. (2017). An integrated participative spatial decision support system for smart energy urban scenarios: A financial and economic approach. *Buildings, 7*(4), 113.

Abastante, F., Lami, I. M., & Mecca, B. (2020). How to revitalise a historic district: A stakeholders-oriented assessment framework of adaptive reuse. In Mondini, G., Oppio, A., Stanghellini, S., & Bottero, M., Abastante, F. (Eds.), *Values and Functions for Future Cities. Green Energy and Technology*. Cham: Springer.

Amistà, R., & Cerreta, M. (2018). Multi-stakeholder spatial decision analysis (M-SSDA) for a culture-led regeneration strategy. In *Computational Science and Its Applications—ICCSA 2018.* Lecture Notes in Computer Science, vol. 10962. Cham: Springer.

Athanasopoulos, K., & Milakis, D. (2014). What about people in cycle network planning? Applying participative multicriteria GIS analysis in the case of the Athens metropolitan cycle network. *Journal of Transport Geography, 35,* 120–129.

Blessi, G. T., Tremblay, D.-G., Sandri, M., & Pilati, T. (2012). New trajectories in urban regeneration processes: Cultural capital as source of human and social capital accumulation—Evidence from the case of Tohu in Montreal. *Cities, 29,* 397–407.

Carius, N., Speck, M., & Laub, K. (2018). *Regional impact assessment: A methodology to measure the regional value added of trans-sectoral urban planning*. Springer, Cham: Green Energy and Technology.

Costa, A. S., Lami, I. M., Greco, S., Figueira, J. R., & Borbinha, J. (2019a). A multiple criteria approach defining cultural adaptive reuse of abandoned buildings. In Huber, S., Geiger, M. J., & de Almeida, A. T. (Eds.), *Multiple criteria decision making and aiding: Cases on models and methods with computer implementations* (pp. 193–220). Springer International Publishing.

Costa, A. S., Lami, I. M., Greco, S., Figueira, J. R., & Borbinha, J. (2019b). Assigning a house for refugees: An application of a multiple criteria nominal classification method. *Operational Research: An International Journal*. Forthcoming.

De Magalhaes, C. (2015). Urban regeneration. *International Encyclopedia of the Social & Behavioral Sciences.*

European Commission. (2014). *Guide to cost—benefit analysis of investment projects: structural funds*. Cohesion Fund and Instrument for Pre—Accession, Directorate—General for Regional Policy, Brussels.

Evans, G. (2005). Measure for measure: Evaluating the evidence of culture's contribution to regeneration. *Urban Studies, 42*(5–6), 959–983.

Ferilli, G., Sacco, P. L., Tavano, Blessi G., & Forbici, S. (2016). Power to the people: When culture works as a social catalyst in urban regeneration processes (and when it does not). *European Planning Studies, 25*(2), 241–258.

Figueira, J., Greco, S., & Ehrgott, M. (2005). *Multiple criteria decision analysis: State of the art surveys*. Boston, Dordrecht, London: Springer Verlag.

Forty, A. (2004). *Words and buildings: A vocabulary of modern architecture*. Londra: Thames & Hudson.

Hoesli, M. (2000). *Property investment: Principles and practice of portfolio management*. Harlow: Longman.

Lami, I. M. (2019) The context of urban renewals as a 'Super-Wicked' problem. In Calabrò, F., Della Spina, L., & Bevilacqua, C. (Eds.), *New metropolitan perspectives. ISHT 2018. Smart Innovation, Systems and Technologies* (vol. 100). Cham: Springer.

Lami, I. M., & Tavella, E. (2019). On the usefulness of soft OR models in decision making: A comparison of Problem Structuring Methods supported and self-organized workshops. *European Journal of Operational Research, 275*(3–16), 1020–1036.

Lami, I. M., Abastante, F., Bottero, M., Masala, E., & Pensa S. (2014). Integrating multicriteria evaluation and data visualization as a problem structuring approach to support territorial transformation projects. *EURO Journal on Decision Processes, 2*, (3–4, 1), 281–312.

Laprise, M., Lufkin, S., & Rey, E. (2015). An indicator system for the assessment of sustainability integrated into the project dynamics of regeneration of disused urban areas. *Building and Environment, 86,* 29–38.

Lazarus, R. J. (2009). Super wicked problems and climate change: restraining the present to liberate the future. *Cornell Law Review, 94*(5), 1153–1234.

Maisonneuve, N., Stevens M., Niessen M., Hanappe P., & Steels L. (2009). Citizen noise pollution monitoring. In *10th Annual international conference on digital government research* (pp. 96–103).

Mak, A., & Stouten, P. (2014). Urban regeneration in Rotterdam: Economic and social values. *European Spatial Research and Policy, 21*(1), 101–122.

McCormack, G. R., Giles-Corti, B., & Bulsara, M. (2008). The relationship between destination proximity, destination mix and physical activity behaviours. *Preventive Medicine, 46,* 33–40.

Metadistretto veneto della Bioedilizia e ITC-CNR. (2012). Strumento di valutazione della sostenibilità energetico ambientale degli edifici—framework, www.comune.rovigo.it.

Miles, S., & Paddison, R. (2005). Introduction: The rise and rise of culture-led urban regeneration. *Urban Studies, 42*(5–6), 833–839.

Millar, R., & Hall, K. (2013). Social return on investment (SROI) and performance measurement. *Public Management Review, 15*(6), 923–941.

Németh, J., & Langhorst, J. (2014). Rethinking urban transformation: Temporary uses for vacant land. *Cities, 40,* 143–150.

Passchier-Vermeer, W., & Passchier, W. F. (2000). Noise exposure and public health. *Environmental Health Perspectives, 108*(Suppl 1), 123–131.

Pensa S., Masala E., & Lami I. M. (2013) Supporting planning processes by the use of dynamic visualisation. In: Geertman S., Toppen F., & Stillwell J. (Eds.), *Planning support systems for sustainable urban development*. Lecture Notes in Geoinformation and Cartography, vol. 195. Berlin, Heidelberg: Springer.

Raviz, S. R. H., Eteghad, A. N., Guardiola, E. U., & Armesto, A. A. (2015). Flexible housing: the role of spatial organization in achieving functional efficiency. *International Journal of Architectural Research, 9*(2), 65–76.

Rittel, H. W. J., & Webber, M. M. (1973). Dilemmas in a general theory of planning. *Policy Sciences, 4,* 155.

Roberts, P. (2000). The evolution, definition and purpose of urban regeneration. In P. Roberts & H. Syke (Eds.), *Urban regeneration, A handbook* (pp. 9–36). London: British Urban Regeneration Association, SAGE Publications.

Salat, S. (Ed.). (2011). *Cities and forms on sustainable urbanism*. Paris: Hermann.
Santangelo, M., Aru, S., & Andrea, P. (2013). *Smart city: Ibridazioni, innovazioni e inerzie nelle città contemporanee*. Torino: Carocci editore press online.
Schneider, T., & Till, J. (2007). *Flexible housing*. Oxford: Architectural Press.
Sepe, M. (2013). Urban history and cultural resources in urban regeneration: A case of creative waterfront renewal. *Planning Perspectives, 28*(4), 595–613.
Tavella, E., & Lami, I. M. (2019). Negotiating perspectives and values through soft OR in the context of urban renewal. *Journal of the Operational Research Society, 70*(1), 136–161.
Tyler, P., Warnock, C., Provins, A., & Lanz, B. (2012). Valuing the benefits of urban regeneration. *Urban Studies, 50*(1), 169–190.
Wang, W. M., Lee, A. H. I., Peng, L. P., & Wu, Z. L. (2012). An integrated decision making model for district revitalization and regeneration project selection. *Decision Support Systems, 54,* 1092–1103.
Yong Yang, A. V., & Diez-Roux, M. D. (2012). Walking distance by trip purpose and population subgroups. *American Journal of Preventive Medicine, 43*(1), 11–19.

Part II
Case Studies

Chapter 6
Theoretical Basis and Design of Analysis

Sara Bonini Baraldi and Carlo Salone

Abstract The chapter highlights the research methodology, which was developed on five different steps. The first stage involved a literature review and focused on three connected key themes (urban regeneration, adaptive reuse, the Italian context). A second stage concerned a reconnaissance of national experiences of recovery and re-functionalisation of under-used or abandoned spaces and the definition of basic criteria to select a limited number of cases to be further investigated. On the basis of the criteria, six cases were selected, which have been studied according to a specific analytical framework developed in phase three. The analytical framework allowed us to identify six different areas of analysis which also reflected the different disciplinary areas of the members of the research group. In phase four, field research has been conducted on the six case studies, collecting data and evidences and through observations, documents and interviews. Phase six consisted in the analysis and interpretative summary with operational indications for decision-makers.

6.1 Introduction

The aim of this research is to study the concrete conditions that can allow the development over time of urban regeneration projects requiring low financial capital, yet which imply high cultural and human capital, through a new approach based on urban, entrepreneurial and cultural projects and new forms of public regulation.

The Italian reality and experiences show how, the traditional 'institutional' conditions—legal rules at local and national scale, bureaucratic routines, decision makers' mindset—represent more of an obstacle than positive premises for the development

S. Bonini Baraldi (✉) · C. Salone
Department of Regional and Urban Studies and Planning (DIST), Politecnico di Torino and Università di Torino, Viale Mattioli 39, 10122 Turin, Italy
e-mail: sara.boninibaraldi@unito.it

C. Salone
e-mail: carlo.salone@unito.it

© Springer Nature Switzerland AG 2020
I. M. Lami (ed.), *Abandoned Buildings in Contemporary Cities: Smart Conditions for Actions*, Smart Innovation, Systems and Technologies 168,
https://doi.org/10.1007/978-3-030-35550-0_6

of adaptive reuse and regeneration practices, in a national context which in fact bears witness to many unused spaces and potentially innovative bottom-up initiatives.

Using representative case studies, this research therefore aims to investigate these conditions in order to verify the possibility of developing new approaches to the reuse of the existing real estate heritage and under-used urban spaces, overcoming the difficulties which have emerged over the past few years and fostering the development of effective, economically sustainable projects and interventions able to generate social, economic and cultural values.

A hybrid method is used, combining different disciplinary approaches—evaluation methods, urban studies approaches, and management science—in a field analysis based on qualitative and quantitative data. The research was conducted according to the following stages:

1. Collection and analysis of bibliographical references
2. Initial reconnaissance of Italian and international cases, processing of selection criteria for cases to investigate, selection of cases to investigate
3. Drafting of the analytical framework and collection of materials on the selected cases
4. Field research for the six selected cases
5. Analysis and interpretative summary with operational indications for decision-makers.

Consistently with the transversal approach of the research, the members of the working group come from different disciplines, thus contributing to the value of an approach embodying many viewpoints on this complex and articulated phenomenon.

6.2 The Key Points of the Research: Urban Regeneration and Adaptive Reuse in the Italian Context

The first stage of the research, involving the analysis, updating and integration of the available literature, focused on three connected key themes, functional to all work stages: a brief reconstruction of the concepts and practices of urban regeneration and adaptive reuse, along with an investigation of the Italian context, seeking to identify emblematic cases.

6.2.1 Urban Regeneration

The macro-field in which the research into Shape, Rules and Values in the contemporary city moves is that of urban regeneration. Adopting this term is clearly exposed to some danger of conceptual abuse and banalisation, but, as we do not have a better way to include in a comprehensive category the experiences collected and investigated, we prefer to take the risk and refer to them as actions in which the urban regeneration is associated to other specific aspects that we will discuss onward.

On one hand, tracing the origins of the vast set of practices which, with some margin of ambiguity and simplification, we define as urban regeneration, is a complex operation which goes beyond the purposes of this chapter. On the other hand, many works in international literature have sought to offer a broad interpretation of this issue, including of course the anthology by Leary and McCarthy (2013) which offers a critical summary of the conceptual framework, describing some cases of the conceptual models, operational traditions and regulatory frameworks in which urban regeneration has been developed.

The ideal pillars on which the urban regeneration experiences rest are essentially two: on one hand, the Enlightenment tradition underlying the Keynesian consensus of the Capitalist west after the Second World War, which preaches the centrality of control over the supply and management of infrastructures of public interest aiming to overcome economic and social inequalities in urban environments; on the other, the increasingly imposing forms of urban entrepreneurialism after the Nineteen Seventies (Harvey 1989) which left increasing margins for the intervention of private capital in the management of local public policies, in line with the neoliberal agenda.

Combined differently over the years, these two elements have laid the regulatory foundations for intervention policies in existing cities, according to a mechanism which has progressively seen the second replacing the first in the practice of urban planning and governance.

In Italy, the late adoption of urban regeneration practices has been contaminated by more traditional approaches, such as those of urban redevelopment, which is much more oriented to the adaptation and renewal of parts of the city which have become obsolete and unsuited, with an increasing awareness towards processes of social and economic change of the urban productive base, developed within a multisectoral dimension (Galuzzi and Vitillo 2018). Thus interventions on the physical aspects (building quality, new functions, equipment), energy-related aspects (standards, materials) and environmental aspects (open spaces, connection to ecological systems and networks) have been joined by other interventions on economic activities and the social structure.

With greater analytical detail, some authors have proposed a reclassification of operations on the constructed city according to intervention methods which are not always clearly distinguished, seeking to create order between the definitions and terms of reference: "*Urban renewal, urban regeneration, urban redevelopment, and urban rehabilitation share similar meanings in the town-planning field but are significantly different in terms of scale*" (see pp. 272–273 in Zhen et al. 2014). *Urban renewal* and *urban regeneration* generally refer to a wider scale: the first indicates physical and structural interventions, the adaptation of structures and spaces in portions of the city which are considered degraded; the second on the other hand is used with reference to complex interventions, aiming to improve or resolve various types of problems, which also affect the environmental, social and economic spheres. On a smaller scale, *urban redevelopment* on the other hand refers to structural interventions focusing on the reuse, with different functions to those originally in place, of one or more buildings, while *urban rehabilitation* indicates the restoration of a building aiming to adapt and relaunch the activities it contains.

To summarise, the expression urban regeneration is used to refer to interests substantially promoted and partly borne by the public sector in urban areas, formally—if not always substantially—responding to the needs of social groups and places suffering from conditions of deprivation (Kleinhans 2004; Leary and McCarthy 2013).

In other words, while urban decline, often concentrated in specific areas of the city in old industrial areas, represents the other face of globalisation which has fostered the spatial concentration of prestigious functions into a few key areas of the world's urban hierarchy and the massive and epoch-making shift of industrial activities to the urban regions of emerging countries, the mechanisms for the regeneration and re-enhancement of the economy in abandoned and impoverished urban contexts have been the dominant contents of urban policies in Western countries over the past thirty years.

The geography of urban abandonment has however different connotations according to the considered region and historical juncture. While some urban areas suffered from such industrial abandonment and radical demographic impoverishment (for a broad critical overview, refer to Pallagst et al. 2013) that they have been referred to as an irreversible process, in others the decline was selective and, in the global cities, tackled with policies that sought to reverse the crisis with huge injections of private capital and intensive real estate enhancement actions (Couch et al. 2008).

More recently, the widespread effects of the financial crisis of 2007 in urban real estate markets in some Western countries led to further lacerations of the material fabrics and in the social make-up of many European urban areas (Knieling and Othengrafen 2015), which rapidly extended to non-urban contexts, causing further forms of demographic and economic impoverishment (for an international analysis see Rodríguez-Pose 2018).

Shape, Rules and Values in the contemporary city deals precisely with these neglected, abandoned and under-used areas: here regeneration is considered the key factor and starting point—or point of arrival—for the re-functionalisation of abandoned buildings, restoring economic and cultural values to the life cycle of urban society.

The contents and tools used to activate these processes see culture, in its broad sense, as a spark and a useful tool. Indeed, under the strong influence of the cultural turn in urban studies inaugurated by the works of Florida (2000), Scott (2000) and many others, the last twenty years have been marked by the dominating centrality of cultural levers in defining regeneration actions, understood both as the material realisation of facilities aiming to create cultural districts in urban centres (Santagata 2002), and an intangible device introduced into (economic and other forms of) value-producing mechanisms (Stern and Seifert 2007). Even, some relate culture and urban regeneration in an indissoluble tie, as strongly stated by Hughes "*My own blunt evaluation of regeneration programmes that don't have a culture component is they won't work*" (Hughes 1998).

More recently, culture and creativity, understood as levers for urban regeneration, have also been related to the issue of social innovation, which can be defined as "innovations that are social both in their ends and in their means [...] new ideas (products, services and models) that simultaneously meet social needs and create new social relationships or collaborations. In other words, they are innovations that are both good for society and enhance society's capacity to act". This enhancement should affect positively "the level of well-being in a community in terms of, for example, education, welfare and social inclusion" (Montanari and Mizzau 2015).

Yet, the academic world and public debate are highly sensitive to these aspects: may works underline the ambiguity of urban policies focused on cultural and creative industries due to their being socially selective and elitist, and very often exploited to justify more or less radical social replacement processes, although there are some experiences in which the presence of local organisations and movements try to maintain a "grass-roots" guide in their initiatives (Rota and Salone 2014).

Even though they do not always focus specifically on cultural issues, many experiences here reconstructed belong to this latter strand, as we will argue in the following sections.

6.2.2 Adaptive Reuse

The reuse of abandoned buildings and spaces in cities represents the heart of the Shape, Rules and Values in the contemporary city research project. The field of reference on this issue is that of the adaptive reuse of buildings, and is part of the urban regeneration context, of which it is one of the key instruments.

Adaptive reuse, as defined by Douglas (2006), refers to a significant change of the function of an existing building or space when its previous function has become obsolete.

According to Robiglio (2016, p. 3) "*reuse happens when individuals or groups introduce new content into an existing container (building, infrastructure, place, area), content that differs from the one for which the container was originally designed*". It is not therefore a container which is adapted to its contents, but vice versa; this allows as much as possible to be conserved, with a radical view of reuse.

Although the expression adaptive reuse is linked to the industrial crisis and is therefore relatively new, the concept itself is not. A theoretical approach to adaptive reuse was established as early as the 19th century when Viollet-le-Duc and Wethered (1875, p. 9) recognised reuse as a way of preserving historic monuments: "*to restore a building is not to preserve it, to repair, or rebuild it; it is to reinstate it in a condition of completeness which could never have existed at any given time*". During the post-war period this concept was lost by the economic boom, but in the second half of the 20th century architects began to consider the possibility of working with historical buildings, initiating debate on reuse from the Seventies.

Moreover, it is possible to distinguish three main approaches to adaptive reuse: *typological, technical and strategic* (Pleovets and Van Cleempoel 2011) Typological reuse distinguishes cases *"according to the building category or building type of the host space"* (see p. 3 in Pleovets and Van Cleempoel 2011). Technical reuse sees *"building adaptation as primarily a technical question. how to adapt a building to allow a new function. In this approach the host space is seen merely as a shell or container and therefore give little attention to the conservation and heritage aspects"* (p. 5). *"The strategic approach focuses on the process and strategies applied for converting significant buildings. Studies classified under the strategic approach generally deal more profoundly with the genius loci than studies that follow the typological or technical approach"* (p. 6).

In our work we make specific reference to the third approach: adaptive reuse goes beyond the mere preservation of the facade of a building in order to satisfy a physical purpose. Adaptive reuse should include the history of a site and its surroundings in order to maximise environmental, social and economic benefits and adapt them to the location. *"Reuse strengthens a community feel by positively linking a city's past to its future, and offering cheap and robust infrastructure to emerging needs, which can spark wholesome renewal processes"* (see p. 6 in Robiglio 2016). The Australian Government Department of Environment and Heritage (2004, p. 3) well expresses this concept, underlining that *"the most successful built heritage adaptive reuse projects are those that best respect and retain the building's heritage significance and add a contemporary layer that provides for the future."*

Adaptive reuse can preserve the character of a city, contributing to the authenticity of the district and the identity of the place, limiting land consumption and improving the general well-being and safety of urban areas.

This practice has also become increasingly popular in relation to global efforts being made towards environmental sustainability and ecological design., The combination of current sustainability standards with existing buildings is a significant challenge in the implementation of adaptive reuse projects which, for this reason, is considered by the majority of authors to be the best alternative, in terms of sustainability, to the construction of new buildings (Douglas 2006; Bullen 2007; Bullen and Love 2010; UNEP 2007; Langston 2008; Conejos et al. 2013; Conejos et al. 2015).

One form of adaptive reuse consists in the temporary reuse of urban spaces or empty buildings, which according to Németh and Langhorst (2014) represents a potential spark for development which sees a leading role by the bottom-up approach, and therefore by the community. The authors acknowledge that urban planning practices have generally moved in the direct of a substantial control of local dynamics and transformations with a typically top-down viewpoint. This approach has caused, or tends to cause, a limitation of the complexity, and therefore the potential for economic and social development of urban fabrics. In this view, a model of temporary use of spaces could be an instrument encouraging more realistic, pragmatic and generative approaches to urban transformation which, considering the time variable, go beyond the paradigm of top-down planning, mirroring the fluidity of communities and local contexts more.

6.2.3 The Italian Context and Cases

We have decided to focus our analysis on Italian cases of abandonment and reuse of buildings given the undeniable proliferation of initiatives structured more or less around this issue that are popping up throughout the country as a new trend in the urban transformations. The growing attention and number of initiatives depends on one hand on the number of disused buildings and spaces available in all the country,[1] and on the other on the practices of reuse spread throughout the different regions,[2] which are often intertwined with social and cultural initiatives, with a clear vocation for youth activities.

There is no database or any complete, updated statistics on disused buildings and spaces in Italy for the time being. Through the initiative "Riusiamo L'Italia", Campagnoli (2014), Campagnoli and Tognetti (2016) provides some data and representative estimates: around 701,978 unused industrial sheds, over 500,000 empty shops, 1700 abandoned stations, at least 20,000 assets available via Fiscal Federalism according to the State Property Office (including over 700 barracks offered free of charge by the Ministry of Defence for 10-year reuse), 56,000 assets available that were seized from the mafia organizations, according to the estimates of Libera, and around 20,000 abandoned buildings of historical/cultural value.

This data offers a measurement of the huge and unexploited heritage available in Italy. This has to be considered alongside the stall—if not decline—of the real estate market, which has also influenced the associated public policies. Therefore, it no longer seems possible to merely simplify the complex set of laws weighing on the public heritage in order to create financial and collective value, while it does seem useful to return to the characteristics of the demand—i.e. those who ultimately use the real estate—to develop policies which offer greater effectiveness in the light of the profoundly changed economic context (Mangialardo and Micelli 2017a, b). Projects and actions of particular importance are represented by forms of bottom-up enhancement, based on initiatives that capitalise on the ability of businesses and third-sector organisations to self-organise.

In Italy, and not only (in this regard see Bailey et al. 2018), the tendency of these actors to use public heritage to experiment practices of social innovation and new entrepreneurship, linked to local communities and resources, is clearly emerging. To cite Mangialardo and Micelli (2017a, b, p. 110) "*bringing this heritage into play means rethinking the underlying role of the public heritage, facilitating the activation of a new economy and new social relations, and consequently, new value*". This new value is also cultivated by the public administrations, promoting—often youth—initiatives which are—or want to be—representative of the so-called creative classe (Florida 2002) and the third sector, able to express innovative ideas and projects. They

---

[1]See for example http://www.spaziindecisi.it/, http://www.disponibile.org/, https://www.wwf.it/il_pianeta/impatti_ambientali/suolo/riutilizziamo_litalia, http://www.temporiuso.org/, https://vuotiaperdereblog.com/.

[2]See for example http://www.riusiamolitalia.it/ita/best_practices.asp, http://www.sudinnovation.it/, http://www.urban-reuse.eu/, http://www.labsus.org/.

do so by making available material resources that would otherwise be unprofitable, but which are in any case able to activate and facilitate the creation of socio-cultural values and new services for their communities. Later on, these practices help to increase the income and real estate value of the spaces used, completing a potentially virtuous circle.

The dynamics that clearly emerge from literature on the Italian context are moving in the direction of meeting the supply of disused (public and private) assets, on one hand, and the demand for services and spaces linked to innovative practices and cultural and social entrepreneurship on the other. This is all part of a field which, at times abused, is defined as social innovation.

Citing Cottino and Zandonai (2012, p. 4) "*Social innovation increasingly represents the guiding concept for those wishing to promote systemic change. It refers to innovation which is social in both its aims and its methods*". This expression describes a set of idea, practices and projects which respond to various kinds of needs and problems, at the same time building new relations and hybrids between sectors and subjects which would otherwise remain independent: in this way they overcome the boundaries of public and private, profit and non-profit, developing new networks and new mechanisms for generating social, economic and cultural values. Ultimately, social innovation represents the synthesis of local needs and the resources, abilities and drive a territory can put into play in order to effectively respond to them.

6.3 Reconnaissance of National and International Experiences, Drafting of Selection Criteria, Selection of Cases for Investigation

Prior to selecting the cases, a reconnaissance was made of national experiences of recovery and re-functionalisation of under-used or abandoned spaces. This first investigation was done by selecting cases which had the following characteristics:

- Single assets in the land registry (initial single unit), with either public or private ownership;
- Assets with no particular attraction in terms of tourism;
- Innovative project within the Italian panorama.

After a careful initial evaluation, the research group decided not to adopt a minimum size threshold for selecting the buildings, considering it more important at this stage to privilege the innovative features over the size of the space.

According to these parameters, 81 cases were identified, out of which 14 in Turin, 14 in Milan, 22 in Northern Italy (apart from Turin and Milan), 5 in central Italy, 13 in Southern Italy. Along with these, 13 European cases were selected for international comparisons (see Appendix 1 for details and Figs. 6.1 and 6.2).

Each of the Italian cases were initially explored using a desk analysis (carried out mainly on the web), in order to classify them according to some standard dimensions

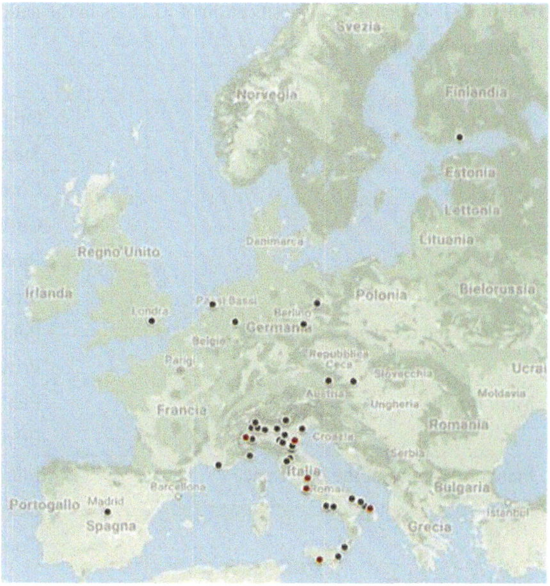

Fig. 6.1 Selected European cases

Fig. 6.2 Selected Italian cases

Table 6.1 Dimensions of classification of 81 cases in the national territory

Name of the experience	
Location	
Type of space	Original function
	Surface area
Type of use	Commercial
	Cultural
	Productive
	Mixed
Origin of the recovery experience	Public
	Private
Main funding	Public
	Private
State of progress of the experience	Initial phase
	Development phase
	Consolidated
Short description	

and provide useful indications for selecting the cases to be studied in more detail. These dimensions are given in Table 6.1.

Following this first census of experiences, it was decided to select a limited number of cases to analyse and investigate further on the field by involving privileged testimonials and researching more in-depth data. The following basic criteria were used for the subsequent selection:

- *Spatial distribution:* it was considered appropriate to select cases from all across the national territory: north, centre, south. This choice derives not only from a need for completeness but is also linked to the need to investigate particularly the relationship between experiences and the normative—and, more broadly speaking, 'institutional'—framework at local level, which has significant differences and specific problems in the various contexts, in which the relations between social stakeholders and the public administration run from cooperation to conflict.
- *Origin of the initiative:* it was deemed appropriate to consider cases based on different origins, such as public initiative, private initiative (not necessarily within a legal framework) and mixed forms of initiative, in order to analyse their different characteristics, strengths and weaknesses, in a comparative logic.
- *Management and governance model,* it was decided to consider public, private and mixed management cases, profit and non-profit projects, in a comparative logic.
- *Current state of progress of the initiative,* it was deemed appropriate to select experiences in different phases of realisation—consolidated, in transition or in crisis—in order to highlight and analyse the main causes and solutions developed by the stakeholders to guarantee their long-term sustainability and effectiveness.

On the basis of the criteria described above, 6 cases were selected: two in the North (Toolbox in Turin and Spazio Grisù in Ferrara), two in the Centre (Officine zero in Rome and CAOS in Terni), and two in the South (Farm Cultural Park in Favara and Ex Fadda in San Vito dei Normanni). The six cases also reflect a fairly representative range of the various dimensions indicated as selection criteria (see Table 6.2).

6.4 Analytical Framework

A specific analytical framework was developed for the case analyses, able to guide researchers in the desk analysis, field research, discussion and drafting of the report. The aim was to outline a clear, exhaustive overview of each case, including basic information such as the physical characteristics, the real estate value, the history of the spaces affected and the reuse project (in architectural, cultural and social terms) involved. Other more specific elements were added, relating to the stakeholders or organisations which began the initiative and their relations, information on the economic and legal management and data on the impacts generated locally and beyond, not only in commercial and market dynamic terms but also in cultural and social terms. This set of information was also ordered within the referred regulatory and legal framework on different administrative levels.

The analytical framework thus allowed us to identify six different areas of analysis which also reflected the different disciplinary areas of the members of the research group (see Table 6.3). The data collected and compared allowed us to identify the strengths and innovative characteristics of each experience of reuse for each area, along with the key obstacles to their development.

(a) *Architectural features and urban context*

The starting point for understanding the case was the study of the physical characteristics of the spaces and structures involved, their history and the events which affected them, but also the urban area they are located in and the urban planning rules governing their development. The analysed aspects in particular focused on:

- Architectural features of the structures and spaces affected: prevailing construction era and architectural typology
- History of the structures and their functions: reconstruction of the sedimentation of uses over time
- Referred urban area: physical characteristics and social and economic context
- Referred urban framework: provision of planning instruments, policies or guidelines for the area by the public administration.

Table 6.2 Summary of selected cases and relative positioning

Case	**Toolbox**
Location	Turin
Origin and management	Private origin, private management for profit
State of progress	Consolidated, growing
Significant aspects for this research	Toolbox is an internationally acknowledged exemplary experience of regeneration and innovation, as it is a completely private initiative. The case has also been studied in terms of entrepreneurial experience, but not in terms of the process of regeneration and reuse of the spaces
Case	**Spazio Grisù**
Location	Ferrara
Origin and management	Mixed origin, private management for profit
State of progress	Consolidated, in transition
Significant aspects for this research	Spazio Grisù is an experience in which the public administration has played a proactive role, on several levels, functional to this private initiative, thus facilitating the construction of an innovative model of asset regeneration and management
Case	**Officine Zero**
Location	Rome
Origin and management	Illegal private origin, private management not for profit
State of progress	In crisis
Significant aspects for this research	Officine Zero is an experience of informal/illegal regeneration of a broad area in which the success of the social and cultural initiatives of the local stakeholders today clashes with the value of the real estate and the urban framework governing it: the bankruptcy auction is held in the green area every year, with the interest of a major banking group which has established a working table with the Officine Zero association to facilitate the reaching of a solution which considers the high social value attained by the asset, however within a real estate operation already at an advanced stage
Case	**Caos Terni**
Location	Terni
Origin and management	Public origin, mixed management
State of progress	In transition/crisis

(continued)

Table 6.2 (continued)

Significant aspects for this research	Caos Terni is an example of real estate operation/initiative of public origin with mixed management on a surface area of 40,000 m^2, which is however currently in a period of instability due to the need to achieve economic sustainability, until now guaranteed mainly by public funds, and marked by conflictual political dynamics
Case	**Farm Cultural Park**
Location	Favara
Origin and management	Private origin, private management not for profit
State of progress	Consolidated, in transition
Significant aspects for this research	Farm Cultural Park is an interesting initiative embracing two main dynamics: the first and most substantial is that of the clash between the owners/managers of the area and the public administration; the second is due to the need for the operation to achieve long-term economic sustainability, which is hard to guarantee today with only private donations. These two serious problems are however contrasted by increasing international public success, and the consequent regeneration of the town centre, with the restructuring of buildings for tourist purposes
Case	**Ex Fadda**
Location	San Vito dei Normanni
Origin and management	Initiative originated from a municipal public competition known as "bollenti spiriti" and the Sandei s.r.l communication firm
State of progress	Advanced, in transition
Significant aspects for this research	The project is recognised as an operation of social corporate responsibility in which, starting from self-building to recover the space, a community of active young people was created and using community economy and culture processes, share a space with access to the training activities and job opportunities

(b) *Origin of the initiative*

Literature on partnership policies aiming to tackle problems of reuse and regeneration is as wide and consolidated as the recourse to public-private consultation in urban requalification and transformation interventions. The scarce effectiveness of public regulatory actions through binding provisions, the continued reduction of public resources to support the action of local authorities and the widespread penetration of neoliberal ideology and the main elements contributing to the codification of consolidated practices of involvement of private interests in urban transformation (Mc Carthy 2008). This is now common practice in the urban landscapes of all latitudes, often even coded as operational planning tools by the relevant legislation,

Table 6.3 Analytical framework for the analysis of the six cases

Analytical dimensions	
Architectural features and urban context	Architectural features of the structures and spaces affected
	History of the structures and their functions
	Referred urban area: physical characteristics and context
	Referred urban framework
Origin of the initiative	Subjects who initiated the project (public, private, mixed)
	Economic resources used to start it and their origin
	Initial aims and project and their development
	Main obstacles met and compliance with the regulations
Governance and organisation	Legal form
	Institutional organisation
	Organizational model
	Stakeholders and networks
Business model and sustainability	Value proposition and public
	Costs and revenues structure
	Critical activities and competences
Impacts on the reference territory	Economic impacts
	Socio cultural impacts
	Impacts on the local real estate market
Market revaluation of the asset	Identification of the value of the spaces involved
	Rental costs of the spaces or rental costs for activities
	Needs and costs of any revaluation works or structure maintenance

which however today tends to be increasingly questioned due to its economic and social effects—segregation of spaces, evictions, social "reclamation"—which often accompany them (Cameron 1992; Smith 2006).

As explained in the following paragraphs, the experiences studied here are rather far from large-scale, high-intensity regeneration processes seen in capitals scattered with many experiences of contemporary urban regeneration. The cases selected are indicative of a bottom-up approach, often promoted by activists who are far from any speculative logics, even though they may recognise the market as their interlocutor and the entrepreneurial approach as their operating logic.

It is fundamental to analyse the project underlying the reuse initiative, with all the organisations, resources and stakeholders mobilised, in order to understand its sense, genesis and future developments, as well as to understand the adjustments and changes required to overcome any obstacles. In this sense, attention is focused on five aspects:

- subjects who initiated the project (public, private, mixed)
- economic resources used to start it and their origin (source, amount of human and financial capital; costs of adaptation and restructuring the spaces and structures)
- relationship with the physical characteristics of the asset (availability of space was the starting point for the design or vice versa? Did the structure and its spaces influence the development of the initiative?)
- initial aims and project and their development (project contents and subsequent modifications)
- main obstacles met and compliance with the regulations (relations with institutions and bureaucratic/normative rules).

(c) *Governance and organisation*

The realisation and long-term sustainability of the experiences depends greatly on the stakeholders involved and how they are organised in order to achieve the results. One part of the analysis was therefore dedicated to the study of organisational and governance models: the most decisive and innovative aspects often emerge from these components.

Governance is an extremely complex concept, which has received much attention from scholars and shows multiple directions of research (Grandori 2000; Gillan 2006; Di Pietra et al. 2008), while never reaching a unique definition (Rhodes 1996; 2007; Windsor 2009; Ruhanen et al. 2010). If blurred borders are typical for social constructs, what increases ambiguity is the use of the term “governance” in two tangential but distinct fields of studies (Ruhanen et al. 2010): political science and corporate management.

Within political sciences, governance refers to “governing with and through networks” or, more broadly, to “a new process of governing; or a changed condition of ordered rule; or the new method by which society is governed” (see p. 1246 in Rhodes 2007). However intended (i.e. with a more or less focus on the self-organisation ability of multiple institutions, Rhodes 1996), here governance is strictly connected to processes of transformation of the role of the State and the emerging role of the private and the voluntary sector (through decentralisation, devolution, privatisation, outsourcing and the like), eradicating its meaning in the widely acknowledged distinction among market, hierarchy and networks. That is, governance in political science refers to how *society* is organised.

A different approach characterises corporate management studies, where governance refers mainly to “the organisation of the relationship between the owners and managers of a corporation” (see p. V in Centre of European Policy Studies 1995) with the objective of protecting the interests of all stakeholders (Bhimani 2008). Notions of accountability and control are stressed, emphasising the role of management and board structure in business contexts (Gillan 2006). Here again the concepts of market, hierarchy and networks come into play, but with a stricter focus on how activities are actually organised in order to minimize transaction costs (Williamson's 1979; Langfield-Smith 2008). That is, governance in corporate management refers to how *business* is organised.

In our research, governance refers to the legal status of the organisation or business unit, the power structure (who are the decision makers and how they are appointed), and how activities are organised. In particular, the areas of analysis included:

- Legal form
- Institutional organisation: structure and composition of the governance, management and control body. Analysis of the involvement of the various stakeholders (province, municipality, companies, other) in the governance and decision-making processes
- Organisational model: number of persons, roles, functions, hierarchical relations, contracts, salaries, professional figures involved, any outsourced activities and functions, with which relations (licences, tenders) and so on
- Stakeholders and networks: analysis of the external subjects mostly involved in the initiative, the cooperation networks activated and the existing forms of influence.

(d) *Business model and sustainability*

Economic sustainability is one of the underlying conditions and at the same time is one of the most critical for the development of regeneration experiences, and for this reason its main sources and components must be analysed: these may be commercial, thus dependent on the public and the users, or rather linked to donations and public or private funding. It is a matter of understanding which business models emerge from the various initiatives.

While the exact definition of what a business model is remains open to discussion (Shafer et al. 2005; Zott et al. 2011), there is widespread agreement on the fact that the concept refers to a representation of how an organisation delivers and monetises the value it creates for its customers (Baden-Fuller and Mangematin 2013; Shafer et al. 2005; Teece 2010). In other words, the model links internal elements like key activities and competences (Osterwalder and Pigneur 2010) to external constituencies like customer segments.

In this research we will focus on the following business model components that we believe are relevant to our analysis. The first relates to *customer group and value proposition* (Baden-Fuller and Mangematin 2013; Osterwalder and Pigneur 2010). Abandoned buildings are used increasingly more often for different purposes by different people: cultural enjoyment, leisure, events, production, hospitality. All these groups have different needs and may be more or less willing to pay for what they consume.

This leads us to the second component of the business model—*revenue streams*—that refers to how and when money is raised. In our analysis we refer both to pricing, timing and methods for collecting revenues (Baden-Fuller and Mangematin 2013), and major costs structures, in order to understand if and how economic sustainability is ensured.

The last element of the business model refers to the *key activities and related key competences*, in order to identify which resources are necessary to activate to create value for the costumers and ensure economic sustainability.

In this sense, the specific elements analysed include:

- Value proposition and public: key mission of the initiative; activities products and services realised and offered; project recipients and type of benefits offered
- Costs and revenues structure: how the various functions activated generate costs and revenues; how the value created for the various beneficiaries translates into inputs (donations, prices paid for services, public contributions etc.)
- Critical competences and resources for the sustainability of the operation: analysis of key resources and competences developed to create value (e.g. availability of certain types of spaces, project management skills etc.).

(e) *Impacts on the reference territory*

Analysing the impacts on the reference territory is useful for understanding the extent to which the project has established strong roots and may also be considered a litmus test for establishing the success of the initiatives, not only in terms of economic performance but also within the social and cultural dynamics of the area. Indeed, one of the work hypotheses was that the analysed experiences and (networks of) organisations could also generate dynamics for transforming the—latent—resources and assets of the territories taken into consideration. The experiments carried out could be stimuli for forms of (individual or group) self-organised private initiatives that are a mix of market, social entrepreneurship and political activism (Salone et al. 2017) The field investigation also aims to highlight the ability of these experiences to stimulate significant relations with other local and non-local stakeholders, creating forms of functional interdependence—local shared spaces and other forms of agglomerative ties—which could improve the effectiveness of these experiences, compensating the lack of financial resources and public planning.

In this sense the analysed aspects include:

- Economic impacts
- Social impacts
- Cultural impacts.

(f) *Market revaluation of the asset*

The fulcrum of regeneration and reuse initiatives lies in the host structures and spaces, which go from being underused or abandoned to constituting a resource and the heart of the activities.

Investigating how the experiences influence their physical surroundings and, above all, how must they affect the increase (or decrease) of market value, is a strategic aspect for developing a new approach, overcoming the challenges which these practices have to tackle today.

The analysed aspects here concerned the market values in the area of the case studied, the restructuring costs, the dynamics of the market, any changes recorded in the market in relation to the presence of the case and the activities performed and so on, including in particular:

The potential effects of the initiatives on the real estate market in the area were investigated:

- Empirically, starting from the interviews with the organisations, in relation to their perception in terms of improvements in the area;
- Scientifically, starting from direct and indirect sources of the real estate market, including Borsinoimmobiliare.com, osservatorioimmobiliarecittàditorino.com;

6.5 Field Research

The field research was conducted between October 2018 and January 2019. For each case, meetings and interviews were held with key figures in the organisation. In order to triangulate the information collected and understand the point of view of the institutions, or the representatives of the individual cases, representatives of the local administrations (particularly planning services) were also interviewed.

All interviews and meetings were recorded and transcribed by one of the researchers. In particular, six people were interviewed for Factory Grisù, for a total of 326 min of recordings; for Farm Cultural Park, five persons were interviewed for a total of 285 min; for Toolbox, the General Director was interviewed for a total of 130 min; for CAOS four persons were interviewed for a total of 205 min; for Officine Zero 15 project activists were interviewed for a total of 150 min of recordings, while for Ex Fadda five persons were interviewed for a total of 148 min of recordings. Parallel to the interviews, the field visits were fundamental of investigating the spaces, activities undertaken and the urban context they are part of, to ensure a full and composite overview of the analysed cases (Table 6.4).

The set of data collected was used to improve the understanding of the realities studied, not only in order to triangulate the data via desk analyses (mainly on line), but also to focus on the analytical dimensions identified in the analytical framework, and on specific aspects planned or foreseen when designing the research. In a necessarily short chapter it is difficult to convey the wealth of contents, projects and unique features of what emerged from the meetings held in the field. These experiences are closely intertwined with the personal stories of the people who brought them to life, and who have invested not only (at times) significant economic resources but also their own personal and relational energies, visions and skills. Chapter 20 attempts to offer a reasoned summary of these situations, and we apologise to those interviewed in advance for any details cut out of their stories.

Table 6.4 List of persons interviewed

Case	Date of interview/meeting	Institution	Person interviewed	Minutes
Factory Grisù	05/10/2018	Consorzio Factory Grisù	Chairman	252
		Consorzio Factory Grisù	Consultant	
		Municipality of Ferrara	Urban Planning Councillor	
		Associazione Spazio Grisù	Founder	
		Tryeco	CEO	39
		Municipality of Ferrara	Culture Councillor	35
Farm Cultural Park	16/11/2018	Ass. Farm Cultural Park	Founder	106
	17/11/2018		Founder	80
	17/11/2018		Worker	26
	16/11/2018	Quid Ristorante	Architect	36
	17/11/2018	Municipality of Favara	Deputy Mayor	37
Toolbox coworking	27/11/2018	Toolbox Srl	General Director	130
CAOS	03/12/2018	Indisciplinarte	Project Manager	90
			Project Coordinator	
		Municipality of Terni	Ex-General Director	75
			Official from the council culture department	40
Officine Zero	14/12/2018	Officine Zero	15 activists occupant	150
Ex Fadda	26/01/2019	Qualcosa di diverso Cooperative	Project Manager	118
			Worker	
			Workers	
			Worker	
		Municipality of San Vito dei Normanni	Urban Planning Councillor	30

Appendix 1: Selected Case Studies

Turin

Case	Link	Topic
Via Baltea 3—Laboratori di Barriera	http://www.viabaltea.it/	Mixed
Toolbox	http://www.toolboxoffice.it/	Coworking
Bunker	http://www.variantebunker.com/	Culture
OGR	http://www.ogrtorino.it/	Culture
Barrio	http://elbarrio.it/	Culture
Rete case del quartiere	http://www.retecasedelquartiere.org/	Social
Hub Cecchi point	http://www.cecchipoint.it	Social
La piazza dei mestieri	http://www.piazzadeimestieri.it	Mixed
La cartiera	http://www.cartiera.org	Social
Open Incet	http://openincet.it/	Innovation
Viadellafucina16	http://viadellafucina16.kaninchenhaus.org/	Art
Copernico garibaldi	https://www.coperni.co/it/torino/copernico-garibaldi	Coworking
EDIT (Incet)	http://www.edit-to.com/about/	Food
Environment park	http://www.envipark.com	Innovation

Milan

Case	Link	Topic
Hangar Bicocca	http://www.hangarbicocca.org/	Art
Base	http://base.milano.it	Mixed
Cariplo factory	http://www.cariplofactory.it/	Incubator
Iulm innovation lab	http://www.iulminnovationlab.com	Incubator
Carroponte	http://www.carroponte.org	Culture
Cascina Cuccagna	http://www.cuccagna.org	Culture

(continued)

(continued)

Case	Link	Topic
Fab Lab	http://www.fablabmilano.it/	Handcraft/innovation
Fucine Vulcano	http://www.fucinevulcano.org	Mixed
Impact Hub	https://milan.impacthub.net	Incubator
Fabbrica del vapore	http://www.fabbricadelvapore.org	Culture
Ex Falck	http://www.milanosesto.it	Mixed
Ass. Cascine Milano	http://www.associazionecascinemilano.org/	Mixed
Macao	http://www.macaomilano.org	Culture
Rimaflow	https://rimaflow.it/index.php/le-nostre-attivita/	Mixed

Northern Italy

Case	Link	City	Topic
Fuoriluogo Asti	http://www.fuoriluogoasti.com/	Asti	Coworking/culture
Diavolo Rosso	http://www.diavolorosso.it/	Asti	Culture
Spazio Kor	http://www.spaziokor.it/	Asti	Culture
Fondazione Pistoletto	http://www.cittadellarte.it/	Biella	Mixed
InStabile	https://www.instabileportazza.it	Bologna	Mixed
Kilowatt	https://kilowatt.bo.it	Bologna	Coworking/ Innovation
Leggere Strutture Art Factory	http://www.leggerestrutture.it	Bologna	Culture
OZ	http://www.ozbologna.org/	Bologna	Culture/sport
FICO	https://www.eatalyworld.it/it/	Bologna	Food
3T PIR	http://www.postindustriale.it	Sellero(BS)	Museum

(continued)

(continued)

Case	Link	City	Topic
Talent Garden	https://brescia.talentgarden.org/		Coworking
Factory Grisù	https://www.factorygrisu.it/	Ferrara	Mixed
Sant'Agnese officina creativa	https://www.santagnese10.it	Modena	Mixed
Casa Bossi	http://www.casabossinovara.com/	Novara	Restoration project
WE Do Fablab	http://www.wedofablab.it/	Novara	Coworking/innovation
Officine ON OFF	http://officineonoff.com	Parma	Coworking/innovation
Filanda Motta	http://www.filandamotta.com	Mogliano Veneto	
Mastronauta	http://www.mastronauta.it	Omegna	Mixed
Progetto Switch	https://www.progettoswitch.it/	Imperia	Mixed
Centrale fies	http://www.centralefies.it	Trento	Mixed
Collezione Maramotti	http://www.collezionemaramotti.org	Reggio Emilia	Museum
M9	http://www.fondazionedivenezia.org/it/attivita/arte-e-cultura/museo-m9	Mestre	Mixed

Central Italy

Case	Link	City	Topic
Impact Hub Firenze	https://florence.impacthub.net	Firenze	Coworking
Le murate	http://www.lemurate.comune.fi.it/lemurate/dove/	Firenze	Mixed
Officine zero	http://ozofficinezero.org/	Roma	Mixed
Casa della pace	https://www.facebook.com/casadellapace.testaccio/	Roma	Culture
CAOS	http://www.caos.museum/	Terni	Culture

Southern Italy

Case	Link	City	Topic
Farm Cultural Park	https://www.farmculturalpark.com/	Favara	Art/social
Exfadda	http://www.exfadda.it/	San Vito dei Normanni	Culture
Zo centro culture contemporanee	http://www.zoculture.it/	Catania	Mixed
Nuovo teatro sanità	http://www.nuovoteatrosanita.it/	Napoli	Art
Albergo diffuso Locorotondo	http://www.sottolecummerse.it	Locorotondo	Tourism
Made in Cloister	http://www.madeincloister.com	Napoli	Handcraft
Fondazione quartieri spagnoli	https://www.foqusnapoli.it/	Napoli	Mixed
Reaction city	https://www.pensandomeridiano.com/reaction	Reggio Calabria	Mixed
Città della scienza	http://www.cittadellascienza.it/	Napoli	Museum
Palazzo innovazione	http://palazzoinnovazione.it/	Salerno	Coworking
Officina San Domenico	https://www.officinasandomenico.it	Andria	Mixed
Officina degli Esordi	http://www.officinadegliesordi.it	Bari	Mixed
Ex Viri	http://www.exviri.it/	Bari	Mixed

International

Case	Link	City	Topic
Tabakfabrik	https://tabakfabrik-linz.at/en/	Linz	Coworking/innovation
NDSM	http://www.ndsm.nl/en/	Amsterdam	Mixed

(continued)

(continued)

Case	Link	City	Topic
kulturbrauerei	http://www.kulturbrauerei.de/	Berlino	Mixed
Holzmarkt	http://www.holzmarkt.com/	Berlino	Mixed
Tate modern	http://www.tate.org.uk	Londra	Art
Spinnerei	http://www.spinnerei.de/	Lipsia	Mixed
Landchaftspark	https://www.landschaftspark.de/	Duisburg	Park
Cable factory	https://www.kaapelitehdas.fi/en/info	Helsinky	Mixed
Friche la belle de mai	http://www.lafriche.org	Marsiglia	Mixed
Matadero	http://www.mataderomadrid.org/	Madrid	Art
Ufa Fabrik	https://www.ufafabrik.de/en	Berlino	Mixed
Westergasfabriek	http://www.westergasfabriek.nl/en/	Amsterdam	Mixed
Wuk	https://www.wuk.at/en	Vienna	Mixed

References

Australian Government Department of Environment and Heritage. (2004). *Adaptive reuse: Preserving our past, building our future*. Canberra, ACT: Commonwealth of Australia.

Baden-Fuller, C., & Mangematin, V. (2013). Business models: A challenging agenda. *Strategic Organization, 11*(4), 418–427.

Bailey, N., Kleinhans, R., & Lindbergh, J. (2018). *An assessment of community-based social enterprises in three European countries*. Power to Change.

Bhimani, A. (2008). Making corporate governance count: The fusion of ethics and economic rationality. *Journal of Management and Governance, 12*(2), 135–147.

Bullen, P. (2007). Adaptive reuse and sustainability of commercial buildings. *Facilities, 25*(1–2), 20–31.

Bullen, P., & Love, P. (2010). The rhetoric of adaptive reuse or reality of demolition: Views from the field. *Cities, 27*(4), 215–224.

Cameron, S. (1992). Housing, gentrification and urban regeneration policies. *Urban studies, 29*(1), 3–14.

Campagnoli, G. (2014). *Riusiamo l'Italia. Da spazi vuoti a start-up culturali e sociali*, Il Sole 24 ORE.

Campagnoli, G., & Tognetti R., (2016). *L'italia da riusare*, La Nuova ecologia.

Conejos, S., Langston, C., & Smith, J. (2015). Enhancing sustainability through designing for adaptive reuse from the outset: A comparison of adaptSTAR and adaptive reuse potential (ARP) models. *Facilities, 33*(9–10), 531–552.

Conejos, S., Langston, C., & Smith, J. (2013). adaptSTAR model: A climate-friendly strategy to promote built environment sustainability. *Habitat International, 37*, 95–103.

Cottino, P., & Zandonai, F. (2012). *Progetti d'impresa sociale come strategie di rigenerazione urbana: spazi e metodi per l'innovazione sociale,* Euricse Working Paper.
Couch, C., Fraser, C., & Percy, S. (Eds.). (2008). *Urban regeneration in Europe*. Wiley & Sons.
Centre of European Policy Studies. (1995). *Corporate governance*. Brussels: Centre of European Policy Studies.
Di Pietra, R., Grambovas, C. A., & Raonic, I. (2008). The effects of board size and 'busy' directors on the market value of Italian companies. *Journal of Management and Governance, 12*(1), 73–91.
Douglas, J. (2006). *Building adaptation* (2nd ed.) Elsevier Ltd.
Florida, R. (2002). *The Rise of the Creative Class: And How It's Transforming Work, Leisure, Community and Everyday Life*. New York: Basic Books.
Galuzzi, P., & Vitillo, P. (2018). *Città contemporanea e rigenerazione urbana*. Temi, azioni, strumenti, Equilibri, Il Mulino.
Gillan, S. L. (2006). Recent developments in corporate governance: An overview. *Journal of Corporate Finance, 12*(3), 381–402.
Grandori, A. (2000). *Organization and economic behaviour*. London: Routledge.
Harvey, D. (1989). From managerialism to entrepreneurialism: The transformation in urban governance in late capitalism. *Geografiska Annaler. Series B, Human Geography, 17*(1), 3–17.
Hughes, R. (1998). *Culture makes communities conference*. Leeds: Joseph Rowntree Foundation, February 13, 1998.
Kleinhans, R. (2004). Social implications of housing diversification in urban renewal: A review of recent literature. *Journal of Housing and the Built Environment, 19,* 367–390.
Knieling, J., & Othengrafen, F. (Eds.). (2015). *Cities in crisis: Socio-spatial impacts of the economic crisis in Southern European cities*. London: Routledge.
Langfield-Smith, K. (2008). Strategic management accounting: how far have we come in 25 years? *Accounting, Auditing & Accountability Journal, 21,* 204–228.
Langston, C. (2008). *The sustainability implications of building adaptive reuse*. Paper presented to CRIOCM2008, Beijing, October/November.
Leary, M. E., & McCarthy, J. P. (2013). *The routledge companion to urban regeneration*. London: Routledge.
Mangialardo, A., & Micelli, E. (2017a). Riuso urbano e immobili pubblici: la valorizzazione del patrimonio bottom up. *Territorio, 79,* 109–117.
Mangialardo, A., & Micelli, E. (2017b). Simulation models to evaluate the value creation of the grass-roots participation in the enhancement of public real-estate assets with evidence from Italy. *Buildings, 7,* 100–113.
Mc Carthy, J. (2008). *Partnership, collaborative planning and urban regeneration*. Aldershot: Ashgate.
Montanari, F., & Mizzau, L., (2015). *Rigenerazione urbana, cultura e innovazione sociale: stato dell'arte e dibattiti in corso*. In *Laboratori urbani*. Organizzare la rigenerazione urbana attraverso la cultura e l'innovazione sociale, Quaderni della Fondazione Giacomo Brodolini.
Németh, J., & Langhorst, J. (2014). Rethinking urban transformation: Temporary uses for vacant land. *Cities, 40,* 143–150.
Osterwalder, A., & Pigneur, Y. (2010). *Business model generation: A handbook for visionaries, game changers, and challengers*. Hoboken: Wiley.
Pallagst, K., Wiechmann, T., & Martinez-Fernandez, C. (Eds.). (2013). *Shrinking cities: international perspectives and policy implications*. London: Routledge.
Pleovets, B., & Van Cleempoel, K. (2011). *Adaptive reuse as a strategy towards conservation of cultural heritage: A literature review*. Conference paper.
Rhodes, R. (1996). The new governance: Governing without government. *Political Studies, 44*(4), 652–667.
Rhodes, R. (2007). Understanding governance: Ten years on. *Organizational Studies, 28*(8), 1243–1264.
Robiglio, M. (2016). The adaptive reuse toolkit. How cities can turn their industrial legacy into infrastructure for innovation and growth. *Urban and Regional Policy Paper, 38,* 5–38.

Rodríguez-Pose, A. (2018). The revenge of the places that don't matter (and what to do about it). *Cambridge Journal of Regions, Economy and Society, 11*(1), 189–209.
Rota, F. S., & Salone, C. (2014). Place-making processes in unconventional cultural practices. The case of Turin's contemporary art festival Paratissima. *Cities, 40,* 90–98.
Ruhanen, et al. (2010). Governance: A review and synthesis of the literature. *Tourism Review, 65*(4), 4–16.
Salone, C., Bonini, Baraldi S., & Pazzola, G. (2017). Cultural production in peripheral urban spaces: Lessons from Barriera, Turin (Italy). *European Planning Studies, 25*(12), 2117–2137.
Santagata, W. (2002). Cultural districts, property rights and sustainable economic growth. *International Journal of Urban and Regional Research, 26*(1), 9–23.
Scott, A. J. (2000). *The cultural economy of cities: essays on the geography of image-producing industries.* Thousand Oaks: Sage.
Shafer, S. M., et al. (2005). The power of business model. *Business Horizons, 48,* 199–207.
Smith, N. (2006). Gentrification generalized: From local anomaly to urban 'regeneration' as global urban strategy. In M. Fischer & G. Downey (Eds.), *Frontiers of capital: Ethnographic reflections on the new economy* (pp. 191–208). Durham: Duke University Press.
Stern, M. J., & Seifert, S. C. (2007). *Culture and urban revitalization: A harvest document.* Philadelphia: University of Pennsylvania.
Teece, D. J. (2010). Business models, business strategy and innovation. *Long Range Planning, 43*(2–3), 172–194.
UNEP. (2007). *Buildings and climate change: Status, challenges and opportunities.* UNEP Publications.
Viollet-le-Duc, E., & Wethered, C. (1875). *On restoration.* London: Sampson Low, Marston. Low and Searle.
Williamson, O. E. (1979). Transaction-cost economics: The governance of contractual relations. *Journal of Law and Economics, 22*(2), 233–261.
Windsor, D. (2009). Tightening corporate governance. *Journal of International Management, 15,* 306–316.
Zhen, H., Shen, G., & Wang, H. (2014). A review of recent studies on sustainable urban renewal. *Habitat International, Elsevier, 41,* 272–279.
Zott, C., Amit, R., & Massa, L. (2011). The business model: Recent developments and future research. *Journal of Management, 37,* 10.

Chapter 7
The Case Study Profiles

Andrea Porta and Francesca Abastante

Abstract The present chapter aims at illustrate the six adaptive reuse case studies basing on the analytical framework's structure described according to the following labels: local context, architectural characters, origin and stakeholders, pills of history and value proposition. This chapter constitutes a synthetic qualitative and quantitative description of the investigated adaptive reuse experiences whose objective is to support and sustain theoretical and practical reasoning about the adaptive reuse concept.

7.1 Introduction

The present chapter aims to illustrate the six adaptive reuse case studies basing on the analytical framework's structure described in the "Theoretical basis and design of analysis" chapter.

Figure 7.1 shows the location of the case studies analyzed in Italy. As it is possible to notice, they are homogenously distributed: two are located in the north of Italy (in Turin and Ferrara), two in the centre (in Rome and Terni) and two in the south (in San Vito dei Normanni and Favara).

This chapter constitutes a synthetic qualitative and quantitative description of the investigated adaptive reuse experiences whose objective is to support and sustain theoretical and practical reasoning about the adaptive reuse concept.

After a brief summary, each initiative is described according to the following labels: local context, architectural characters, origin and stakeholders, pills of history and value proposition. It is important to underline that the information collected for

A. Porta · F. Abastante (✉)
Department of Regional and Urban Studies and Planning (DIST), Politecnico di Torino,
Viale Mattioli 39, 10122 Turin, Italy
e-mail: francesca.abastante@polito.it

A. Porta
e-mail: portandrea.me@gmail.com

© Springer Nature Switzerland AG 2020
I. M. Lami (ed.), *Abandoned Buildings in Contemporary Cities: Smart Conditions for Actions*, Smart Innovation, Systems and Technologies 168,
https://doi.org/10.1007/978-3-030-35550-0_7

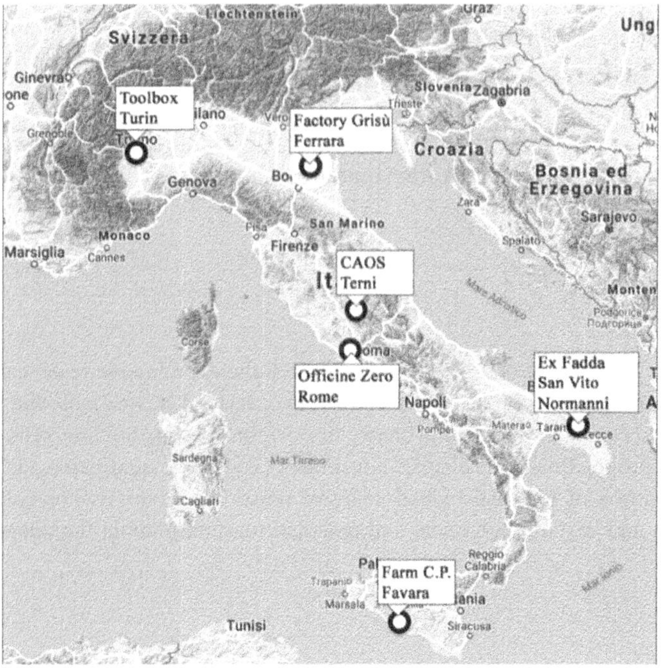

Fig. 7.1 The six Italian case studies. *Source* Own elaboration

each case study are mainly based on local or national press as well as on direct interviews to the stakeholders involved in the different adaptive reuse processes. This causes an inhomogeneity of the available information.

7.2 Toolbox Coworking

7.2.1 In Brief

Toolbox (toolboxoffice.it) is a hub and a space for coworking and innovation. It is located in Turin, in the buildings that used to be a fashion company and a foundry.

This entrepreneurial initiative was developed by the business group owner of the spaces and of the "GB Sportelli" fashion company, that was located there before its closure and the development of Toolbox.

The structure houses numerous offices and work stations available for freelance professionals and firms from various sectors, offering services, spaces, workshops and events related in particular to culture, technology and innovation.

Fig. 7.2 The Toolbox coworking area

7.2.2 *Local Context*

The industrial area in which today stands the Toolbox Coworking is called "OSI GHIA Area" and covers about 51,000 m^2 of land. It is located in Turin, in the so called "Crocetta" district, around 1500 m from the Porta Nuova train station. The "OSI GHIA Area" is wedged between the South railway branch (towards Genoa, Italy), the North railway branch (towards Milan, Italy) and the Corso Dante railway bridge (Fig. 7.2).

At the beginning of the 20th century, the "OSI GHIA Area" was located outside the residential districts of Turin, in an area that today we would define as the outskirts where companies used to establish their numerous productive activities (Balestra and Ferrero 2016).

Nowadays, this area is formally part of the "Crocetta" district. However, due to its specific location, it is still isolated from the district. In fact, the conformation of the area, closed between the railway lines and the Course, makes it rather isolated, contributing to the state of abandonment and degradation of the industrial buildings (Toolbox project manager, interview 27/11/2018).

In 2008, the "OSI GHIA Area" has been subject to an Executive Plan (EP), that provided a change of destined use from industrial to residential, commercial and public. The idea was to destine a portion of the area to a university campus and a portion to residences and shops. The EP, due to the economic crisis of 2009, was abandoned (toguardainalto.altervista.org; Balestra and Ferrero 2016).

7.2.3 *Architectural Characters*

The "OSI GHIA Area" occupies about 51,000 m^2 of land and within it stands the Toolbox main building for a total of 5872 m^2 of covered surface.

The main building has been designed by the Arch. Porcheddu for the Carlo Garrone Foundry.

The prevailing material is reinforced concrete. The structure of this main building has remained substantially unaltered over the years. This is due to a constraint imposed by the Municipality of Turin devoted to protect the building for its value of industrial archaeology. A part of the main building, the Toolbox coworking currently occupies two other buildings. The overall distribution of the buildings can be described as follows (Balestra and Ferrero 2016):

- two buildings of three floors above ground and a basement constituting two sleeves. Each floor is about 750 m^2 with a structure “insoles and pillars” in reinforced concrete. This structure turned out to be very useful since it allows the realization of big open spaces and windows;
- a main building of three floors above ground of about 750 m^2. This building used to contain the offices and the personal apartment of the director of the former foundry. The structure is identical to the two other buildings. However, currently this building still shows the original distribution of the spaces and windows;
- a “plate” on the ground floor, which is a covered courtyard between the two sleeves and the main building. The plain roof of the covered courtyard is currently a parking sustained by a network of square mesh pillars under which there is 3000 m^2 open space destined to warehouse (Fig. 7.3).

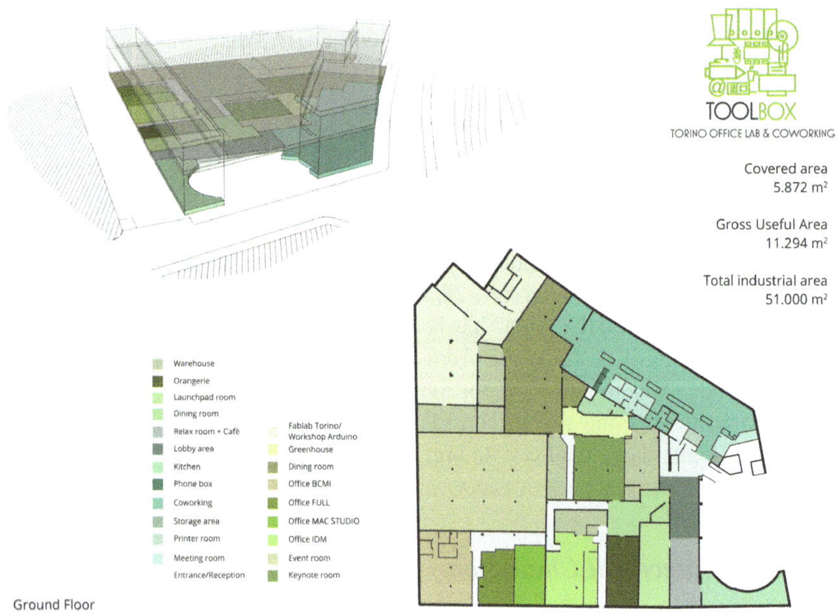

Fig. 7.3 Toolbox scheme

The first renovation required to make the Toolbox coworking came true, involved the first part of the main building and the plate. It was made by maintaining the original structures, treated as a casing, inside which were inserted the new spaces, characterized by a strong design component (divisare.herokuapp.com; tiazzoldi.com/m-toolbox).

7.2.4 *Origins and Stakeholders*

Toolbox is a private operation since the beginning. It was developed by the business group owner of the buildings that had to define new functions for the spaces in order to minimize the economic loss derived from the closure of the former fashion company (Toolbox project manager, personal interview 27/11/2018).

The team that was entrusted with this operation was composed by three main figures: (i) a representative of the business group property, responsible to vet and endorse the final decisions; (ii) a project leader, which was the former manager of the fashion company and responsible for the role of ideation and management; (iii) an architect called to design the overall spatial concept (Balestra and Ferrero 2016).

According to the sources (presstletter.com 25/03/2011) the needed initial financial investment was of 500,000 €. That amount of money was paid by the business group property in order to refurbish a portion of the main building, around 1500 m^2, and to open the Toolbox coworking.

7.2.4.1 Pills of History

1919: The Foundry Garrone was designed by the Arch. Porcheddu;

1954: The GHIA bodywork has been opened in a portion of the area;

1960: The "Officina Stampaggi Industriali" (OSI) has been opened occupying the spaces next to the GHIA bodywork. From now on, the overall area is called "OSI GHIA Area";

1968: The Foundry Garrone has been closed;

1978: The "GB Sportelli", a very well-known fashion brand, opened in the buildings that used to be of the Foundry Garrone;

2000: The overall OSI–GHIA complex has been closed. However, the "GB Sportelli" continued to maintain its activities on the area;

2007: The OSI–GHIA complex has been acquired by an investment fund with the aim of building a new residential and commercial area (toguardainalto.altervista.org 2008) while the "GB Sportelli" continued to maintain its activities;

2008: the Municipality of Turin approved a Convention Executive Plan (CEP) providing new destinations for the area as follows: 25,000 m^2 as residential, 15,000 m^2 as commercial, 10,000 m^2 service activities (professionearchitetto.it 11/05/2011; Balestra and Ferrero 2016). This CEP gave birth to a GRP variation in 2011, that has been abandoned due to the economic crises occurred in 2008 (Toolbox project manager, interview 27/11/2018).

2008: The "GB Sportelli" closed its activities because of business reasons but the business group maintained the ownership of the buildings.

As affirmed by the Toolbox project manager (personal interview 27/11/2018) the closing of "GB Sportelli" determined the necessity of the property to identify a new life for the building, which is located in a strategic but difficult position in the centre of the industrial complex. Accordingly, the selling of the building and the possibility of placing new productive activities turned out to be very difficult also due to the new born economic crisis;

2009: The Toolbox coworking project was born. Looking at the USA experiences, the former manager of the "GB Sportelli" proposed to create something new for Italy and Turin: a coworking space destined to independent professionals. According to their needs, the property decided to finance the project (Balestra and Ferrero 2016).

2010 (8 April): The Toolbox Coworking opened occupying 1500 m^2 GLA;

2011–2015: The Toolbox team enlarged the project recovering other spaces for a total of 6000 m^2 on 10,000 total square meters (ricerca.repubblica.it 6/12/2016).

7.2.5 Value Proposition

As in the beginning, the Toolbox coworking is currently a private initiative with profit purposes. Its business is mainly based on renting coworking spaces and services for a total amount of about 400 users and 150 activities (Project manager, personal interview, 27/11/2018).

The professionals and freelancers that want to settle their activities in the Toolbox spaces need to pay a rent. In exchange, the Toolbox coworking offers a private/shared space and a number of services as: internet fibre, reception services, access 24 h, conference rooms, kitchens, a creative environment for private events and a great number of conferences.

In addition to the work stations and shared offices, Toolbox hosts a fab-lab (an association dedicated to printing techniques) and an experimental smart house (toolboxoffice.it).

7.3 Factory Grisù, Ferrara

7.3.1 In Brief

Factory Grisù (factorygrisu.it) is a creative hub and business located in Ferrara (Italy) in a building that used to be a fire station. The owner of the building is the Province of Ferrara that gave it to the Municipality of Ferrara as free loan for use. Currently, the spaces are managed by a consortium of firms called Factory Grisù, whose components live and work inside the building.

The activities promoted by the Factory Grisù are related to cultural and creative for-profit industries. However, it deals also with the development of social projects and initiatives for the district and the city. In terms of costs, each firm of the consortium does not pay any rent to the Municipality, but they have to restructure and maintain the spaces occupied.

7.3.2 Local Context

Ferrara is the provincial capital of the Emilia Romagna Region (Italy), counting for 132,278 inhabitants (ISTAT 2018).

The Factory Grisù initiative is located within the historic city walls, in the so-called "Giardino" district (Fig. 7.3). This district is part of the city centre of Ferrara that is built on the foundations of the seventeenth-century pontifical fortress and it very well-known in Italy for its architectural values (Guzzon 2012).

According to the General Regulatory Plan (GRP) of the Municipality of Ferrara (1995) the former fire station used to host functions of leadership. In 2008, the Special Retraining Programme called "Actions for the urban development of the areas of excellence of the city of Ferrara" (i.e. *Azioni per lo sviluppo urbanistico delle aree di eccellenza della città di Ferrara*), established to dismiss the complex of buildings, demolishing part of them and reconstructing other buildings proposing new functions. As affirmed by the Councillor for town planning of Ferrara (personal interview 5/10/2018) although approved, this project has never been developed.

According to Monti (2010), in the last ten years the "Giardino" district experienced a period of depopulation and criminality that led it to be considered and perceived as marginal and problematic by the local population (estense.com). Despite the negative reputation of the district, it is possible to recognize different accommodation and restoration activities supported by the proximity of the railway station as well as the presence of the "Meazza" stadium and services (Fig. 7.4).

7.3.3 Architectural Characters

The overall area hosting the Factory Grisù is about 3600 m^2 of which 1565 m^2 are covered.

The former fire station building has been designed by Ing. Luigi Barbantini and opened in 1930 (Monti 2010). This building is one of the architectural elements that contributed to redesign the "Giardino" district characterizing it in terms of services and tertiary.

As described by Monti (2010), the original main body consists of a single building, with two floors above ground, in addition to a manoeuvring tower. However, during

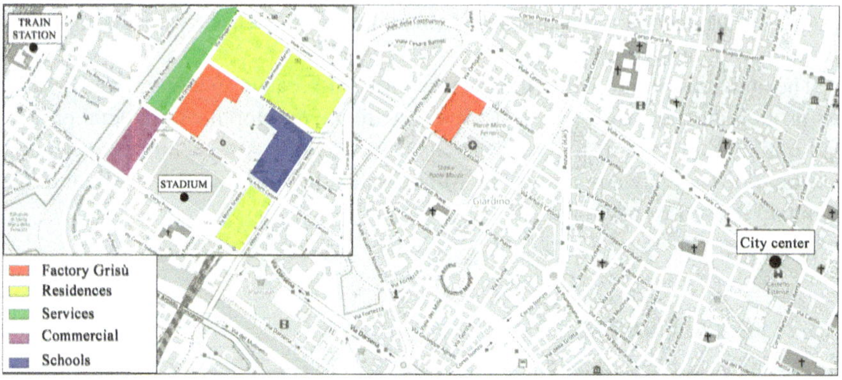

Fig. 7.4 The "Giardino" distric

the years, the building has been subject to various enlargement and renovation works making it being currently characterized by wings surrounding an internal courtyard, apart from the original structure (Fig. 7.5).

More in details, the north-west and north-east wings have a rectangular plan with a 2-pitch roof. The square-shaped tower and flat roof terrace is crowned at the top by a masonry balustrade set on a pronounced cantilever cornice. The corner portion is also rectangular in plan and is joined to the side wings with geometric compensation

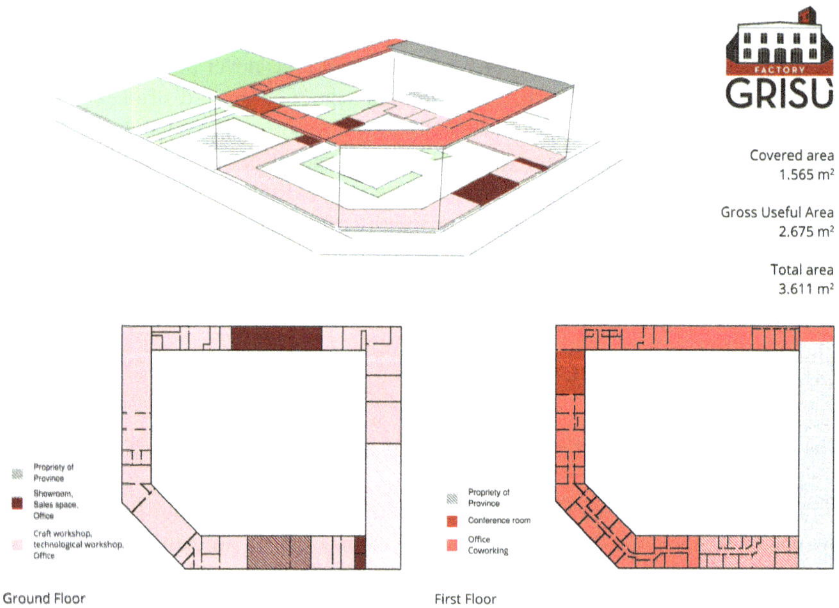

Fig. 7.5 Factory Grisù scheme

volumes mainly used for stairwells. The south-west wing, built between 1949 and 1953, consists of a rectangular unadorned building, a gabled roof, a single floor to roof functional for the use vehicles. Finally, the south-east wing, built between 1979 and 1982, consists of an unadorned building with a rectangular plan, a gable roof, two floors above ground.

The building maintained his original destination of fire station until 2004, when has been dismissed remaining intact from a structural point of view even if in a bad state of maintenance.

The roof, the courtyard, the common spaces and the basic installations have been restored by the Province of Ferrara (the works are still ongoing). According to President of the Factory Grisù (personal interview, 05/10/2018), more than 50% of the overall spaces of the former fire station has been occupied by the Factory Grisù. Those spaces have been restored directly by the firm's participant to the consortium. This implied that the works have been done in a rough way mainly due to the temporary nature of the contracts.

7.3.4 *Origins and Stakeholders*

The Factory Grisù initiative originated by the private Grisù Association, founded *ad hoc* in 2012 to develop the project "Spazio Grisù" (spaziogrisu.it), with the proactive support of the Province and the Municipality of Ferrara (Agnoli 2014; Mangialardo and Micelli 2017). The Grisù Association was composed by local activists, committed in art and culture sector. At the beginning, no money was invested in the project, but only volunteering and work done directly by the Association members in order to temporary reuse the spaces with cultural and social functions, as illustrated by one of the activists interviewed (personal interview 5/10/2018).

7.3.4.1 Pills of History

1930 (28 October): inauguration of the fire station;

1943: The province of Ferrara became the owner of the building that maintained its original use of fire station (Guzzon 2012);

2004: The fire station has been dismissed due to the need for new available spaces and the firefighters were transferred to new barracks. The building has been abandoned, becoming warehouse of the Province of Ferrara (President of the Factory Grisù consortium, personal interview 5/10/2018);

2010: The Province of Ferrara auctioned the building for a base of 3.5 million € (Monti 2010) but, as stressed by the Councillor for town planning (personal interview 5/10/2018) nobody was interested in buying it);

2011: The Province of Ferrara re-auctioned the building, lowering the auction base to 2.8 million € but again nobody was interested in buying it (Agnoli 2014, 2015);

2012 (22 March): The Grisù Private Association requested the temporary use of the structure to the Province with the "Spazio Grisù" project. The association, composed by local people, with artistic, cultural and social engagement, was set up in order to revitalize the unsold fire station making a cultural hub (Agnoli 2014, 2015);

2012 (7 September): The Province, in agreement with the Municipality of Ferrara, accepted the "Spazio Grisù" project and gave for free the barracks to the Grisù Association for 5 years. The first idea was of a temporary use. In this sense, if anyone was interested in buying, the Province could have ended the "Spazio Grisù" experience (Councillor for town planning, personal interview 5/10/2018).

The idea was that no public money or resources would be used or asked: the restoration of the building would be implemented and payed by the people and activities settled in the spaces (Agnoli 2014, 2015). During the first months the members of the Grisù Association roughly cleaned up the place with the informal help of local citizens.

The role of the Association was to coordinate the project as a whole and to select the activities. They were not settled in the spaces of the barracks and their objectives were strictly no-profit oriented (spaziogrisu.it);

2012 (19 October): the Grisù Association launched the first call for the establishment of businesses and experiences in order to attract creative and cultural activities (Agnoli 2015; spaziogrisù.it).

2013 (21 March): the Association gave the spaces to 13 companies and associations selected. Each of them started the adaptation and the restore of the spaces according to the needs of the activities. Due to the temporary nature of the initiative, the restructuring works were done in a rough and simple way, in order to facilitate the return of the investments in the five years of the concession (Agnoli 2014, 2015). At that time, the current President of the Factory Grisù consortium was one of the association's members and a component of an activity established in the "Spazio Grisù".

When interviewed (personal interview, 5/10/2018) he affirmed that, from the very beginning it was noted that it was very difficult for the established experiences to be able to restore and adapt all the common areas. This difficulty contributed to determine the failure of the idea of being autonomous from an economic point of view;

2013: The Province of Ferrara applied a regional call for support for competitiveness and required a contribution of 800 thousand € to restructure the barracks in order to guarantee the future of the "Spazio Grisù" initiative responding to the structural needs mainly related to common spaces, plants and roofs (as confirmed by the President of the consortium and the Councillor for town planning in the interview of 5/10/2018);

2014: The Region financed the 50% of the requested contribution (i.e. 400 thousand €) while the remaining amount was paid by the Province (estense.com 29/03/2014).

In the same year, following the Italian Legislative Decree n.56/2014 (DLG 7/04/2014) that limits the powers of the Provinces, the Province of Ferrara entrusted the property on loan for use to the Municipality of Ferrara in order to continue the project. According to the Councillor for town planning and the president of the consortium (personal interview 5/10/2018), these normative changes determined a slowdown of the restructuring process obliging the Province, the Municipality and the Association to rethink the management scheme in relation with the necessity of using the Region funds.

Indeed, the Grisù Association was not fitted, in normative terms, to manage the operation and guarantee the characteristics requested by the Region, but also to answer the critics coming from the citizenship, emerging on the local media and press, regarding the lack of transparency of the experience and the use of public funds (estense.com 29/03/2014). After some attempts and discussions regarding the possibility of adaptation of the association, the solution established was to re-orient the project in a for-profit purpose, launching a new call for the management of the fire station, excluding the possibility of application for the association, that decided for its closure;

2016: The Grisù Association changed its legal status becoming "Factory Grisù consortium";

2016 (14 June): The Municipality of Ferrara launched a procedure of public evidence for the management of the property and the renewal of the "Spazio Grisù" project which was assigned to the new-born Factory Grisù Consortium until March 2025.

The Consortium was founded by nine members of nine firms already established in the fire station who are committed to developing the project with its own economic, managerial and labour investments (factorygrisù.it). From this time on, the firms can establish their activities in the "Factory Grisù" project until 2025, after a selection procedure cyclically managed by the Consortium and the Municipality of Ferrara. To do this, the firms do not pay any rent but a *quota* of 1300 € is foreseen to join the consortium plus 300 € of contribution.

7.3.5 Value Proposition

Currently, the Factory Grisù Consortium is a profit organization, which is why it deals first of all with the development of the companies that are part of it.

As stressed by the president of the consortium (personal interview 5/10/2018), the objective of the Consortium is related to the regeneration of the fire station. Moreover, it aims at enlarging the number of firms, becoming a big player of the cultural and creative sector, capable of applying to calls at the European level.

The idea is to create a large community of firms and professionals in the field of culture and creativity, covering every part of the supply chain, working together in upgrading the business.

As emerges in the call designed by the consortium (Factory Grisù 2016), today the firms benefit from the services provided by the consortium and its partner/contracted suppliers: administrative and accounting services; support for the design of tenders and business plans; insurance services; fibre internet connection; digital communication services; basic supplies: energy, heating, cleaning, sharing the management costs.

Each company is required to participate in the development of the project by joining the Consortium and participating in the economic and operational commitments that it has contractually assumed with the Municipality of Ferrara. In particular, the assignee companies assume the costs of completing the works necessary to make the assigned premises suitable for their activities.

In addition to the Factory Grisù consortium deals with organizing initiatives, events and projects of a social and cultural nature in favour of the development of the district and the city in which it is located.

7.4 Centro Arti Opificio SIRI

7.4.1 In Brief

The Centro Arti Opificio Siri (CAOS, caos.museum.it), is a centre of production and fruition of arts and culture, conceived and owned by the Municipality of Terni, as a portion of the recovery of the former "SIRI" industrial area (i.e. *Società Italiana Ricerche Industriali*).

Till 2020, CAOS is managed by a temporary association of companies called "Indisciplinarte Srl", in the context of a public private partnership, regulated by a 5 years public funded management tender. CAOS involves both permanent and temporary exhibitions and activities as: permanent the Museums of Modern and Contemporary Art Aurelio De Felice and the Archaeological museum, the "Sergio Secci" Theatre, the Area-Lab dedicated to educational and creative workshops, ateliers for residences and artistic productions, a specialized library, a video room and a coffee restaurant as well as an area dedicated to national and international temporary exhibitions.

7.4.2 Local Context

Terni is the capital province of the Umbrian region and counts for about 110,000 inhabitants (ISTAT 2018). This city is very well known in Italy due to the industrial development in the iron sector that characterized this territory until the nineties. After the financial crises occurred in 2008, many industries started to be dismissed resulting in a huge number of abandoned buildings (Tarquini 2002) (Fig. 7.6).

The CAOS initiatives is located in the so called "Città Giardino" district located in a strategic position next to the city centre and in an archaeological Roman area showing numerous prestigious buildings (Sposito 2012; Tarquini 2002). The district

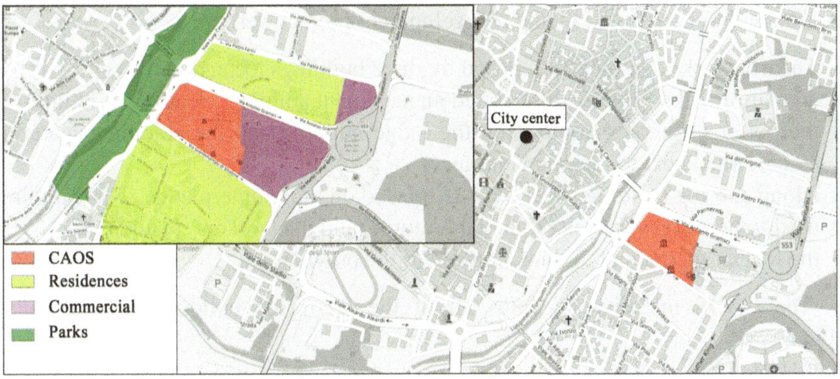

Fig. 7.6 The "Città Giardino" district

and, in particular, the specific area in which CAOS is place, have public and cultural destination according to the GRP. However, the area has been subject to numerous urbanistic tools such as a Programme of Urban Renewal (PUR) and a Urban Complex Programme (UCP) promoted and developed by the Municipality of Terni with the aim of restoring the area with cultural functions (Tarquini 2002; Sposito 2012).

With the opening of CAOS (2010) the district is the subject of a redevelopment project in which it assumes the role of creative district (Former Councillor for town planning personal interview 3/12/2018; Umbria.it 24 16/11/2013; Municipality of Terni 2016).

7.4.3 Architectural Characters

The CAOS occupies around 44,000 m^2 of land of a former iron work industrial area of 44,000 m^2, characterized by a high quality in terms of landscape and architectural values. The area is accessible through a tree-lined avenue which ends in a pedestrian courtyard constituting the centre of the original industrial area's nucleus (Sposito 2012; Tarquini 2002). As a inheritance of the former industry, it is still present a channel of derivation of the river waters for the production of electricity.

The original redevelopment projects provided the overall recovery of the pre-existing buildings (Public representative in charge of the redevelopment project, personal interview 3/12/2018). In order to preserve the architectural industrial characters of the buildings, it was decided to make conservative restoration interventions aimed at functional and formal recovery. In fact, the Public Administration (PA) affirmed that the area at stake constituted an important testimony of the first industrial settlement of Terni (Tarquini 2002).

The buildings affected by the redevelopment project were:

- The ironworks complex, composed by three buildings located in the centre of the area (ironworks building, offices and management building, electrolysis building);
- the porter's complex comprising the lodge building and the adjoining material storage shed next to the main entrance;
- a long and big building shaped as "L" placed on the southern edge. This building used to host the warehouses, the laboratories and the workers' residences;
- the former methanol building annexed to the previous.

The CAOS initiative occupies all of the recovered buildings, in particular (Fig. 7.7):

- the central complex hosts the civic museum of modern urban history and industrial archaeology of Terni, an art gallery, multi-purpose spaces for projections, exhibition activities, conferences and training;
- the porter's complex hosts an information centre and guest house dedicated to hospitality under construction (under construction);
- the "L" building hosts the archaeological and paleontological sections of the civic museum, a library and documentation centre on the history of the site, a café.
- the methanol building hosts a theatre of about 370 seats.

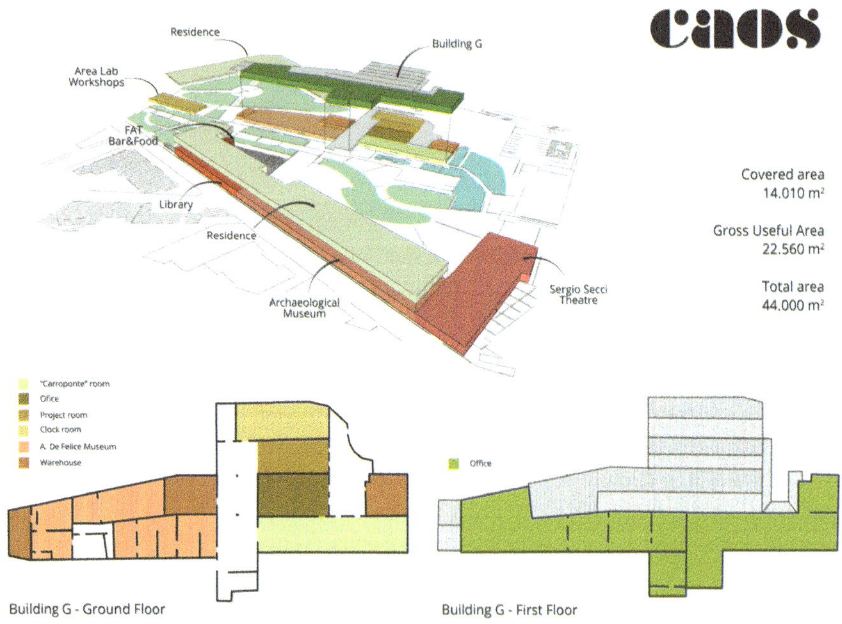

Fig. 7.7 The CAOS scheme

The new part, built in the framework of the project promoted by the municipality of Terni for the whole area, concerns a large commercial space with underground garage parking and basement, and a share of residences.

7.4.4 Origins and Stakeholders

The project was developed and managed by the Municipality of Terni as part of the redevelopment path of the industrial area of the former SIRI, with the objective of converting it with a cultural function, preserving the historical parts.

The project was carried out in partnership with private companies interested in investing in the area (Tarquini 2002; Sposito 2012). The overall public-private investment (considering the commercial parts) was about 65 millions €, of which about 8 millions were dedicated to the building of CAOS (Sposito 2012; Civita.it 28/03/2009).

The configuration of CAOS as a centre of contemporary art, with a focus on production and events, according to the municipality functionaries interviewed (personal interview 3/12/2018) was born thanks to the input of the "Indisciplinarte Association", composed by some of the people today dedicated to the management of the initiative. The association in fact, according to the project manager (personal interview 3/12/2018) in 2006, when the renewal operation was still ongoing, proposed a temporary use of the buildings for an art festival.

7.4.4.1 Pills of History

1793: construction of the Pontifical State factory for ironwork;

1925: The factory became a chemical industry of the Italian Society for Industrial Research (SIRI);

1985: The SIRI closed (Sposito 2012) and the buildings have been dismissed. Accordingly, the private owner proposed to transform the area in order to re-activate the industrial production with new porpuses. However, this project was not approved by the Municipality of Terni that preferred to include the former SIRI area into the new GRP as an historical area (Former Councillor for town planning personal interview 3/12/2018);

1994: The Municipality of Terni acquired the central historical buildings (ironworks complex) after numerous interactions and bargains with the property. The instrument used to develop the operation was the PUR that gave the money for the acquisition and first restoration of the buildings (Tarquini 2002; Sposito 2012);

1997–2002: The Municipality of Terni, in partnership with a private investor (Coop Umbria casa) interested in building residencies in part of the area, continued the restoration and reuse project of the whole SIRI area, included the acquisition of the other parts (the "L" building), through the application of a UCP approved by the Umbria Region. In that years the cultural destination (museums and libray in particular) was planned. (Tarquini 2002);

2002–2009: The great part of the open area surrounding the industrial buildings owned by the municipality was still private and abandoned, so the Municipality had the idea of involving another private investor interested in building a mall in Terni (personal interview 3/12/2018);

The project was to help the investor Coop Centro Italia, from the urbanistic point of view in order to facilitate the settlement and building of the new supermarket in the area. In this way the new investor financed the works needed for the urbanization: paths, roads, main square, garden.

2004: Opening of the Archaeological museum and library in the "L" building, that was originally dedicated to warehouses and worker's residencies. The other buildings were still under restoration.

2006: A small share of the central complex, newly restored and waiting for a destination, was given for temporary use to the "Indisciplinarte Association" composed by local citizens engaged in the culture sector. The "Indisciplinarte Association" organized a successful arts festival.

2008–2009: After the temporary use experience, having decided to dedicate the whole buildings to create a cultural hub, the Municipality of Terni implemented a call for its management. The call, in 2009, included about 3 million € for 5 years (civita.it 28/03/2009; umbria24.it 18/12/2013).

The call was won by a temporary association of companies led, for the management part, by "Indisciplinarte Association" (which became a cultural enterprise after the temporary use experience), with two social cooperatives, and CIVITA, a national player in the field of culture and museums.

2009 (28 March): The CAOS was open (civita.it 28/03/2009).

Over the years, CAOS and its management team were at the centre of a huge political and media controversy regarding the amount of funds, the private involvement, the limited number of visitors and the contents of the events and exhibitions. This argument has been widely reported by the local press as umbria24.it (18/12/2013; 8/08/2014);

2014: The Municipality of Terni launched a second call for the five years management of the CAOS area. "Indisciplinarte Association" won the competition acquiring 2 million € of available funds. Again, this call and "Indisciplinarte Association" have been in the eye of the storm due to political controversies. This argument has been widely reported by the local press as lanazione.it (8/10/2014), umbria24.it (2/3/2017), umbria24.it (24/05/2017);

2018: The management contract is about to expire in October 2019. The new Municipality of Terni (elected in 2016) is strongly expressing the willingness to rethinking the management model, cutting the available funds for CAOS (Terni Today.it 3/09/2018).

7.4.5 *Value Proposition*

The main mission of CAOS is to promote and disseminate different forms of culture and arts. In fact, together with museums and exhibitions, the CAOS has a calendar of cultural events and activities dedicated to citizens and to primary and secondary schools' education (caos.museum).

CAOS is currently composed by (caos.museum.it, personal interview 3/12/18):

- The Aurelio De Felice Museum of Modern and Contemporary Art, born from the fusion between the Municipal Art Gallery and a specially created collection dedicated to contemporary art, is made up of 4 main areas (antique painting, monograph by Aurelio De Felice and Orneore Metelli, collection of graphics, contemporary section);
- the Archaeological Museum of Terni which is divided into a pre-roman and a roman section;
- a library in which numerous texts of art, archaeology, history and local publications are collected;
- temporary exhibitions area;
- the AreaLab dedicated to educational and creative laboratories and local associations;
- the "Sergio Secci" Theater;
- atelier for residences and artistic productions;
- the Fab Lab of Terni managed by the HackLab association;
- video Room;
- a coffee restaurant;
- a guest house and space dedicated to hospitality under construction;
- CAOS promotes and organizes various cultural events and festivals.

7.5 Officine Zero, Rome

7.5.1 *In Brief*

Officine Zero (OZ, ozofficinezero.org) is an illegal initiative proposed by collective of workers and freelancers with social and political objectives. It was born in 2012 after the illegal occupation of workers in layoffs and local activists of the former Rail Service Italia (RSI) buildings in Rome, next to the Tiburtina Train Station.

The overall area occupied by Oz Officine Zero is of around 29,000 m^2 of land. After the failure of the part involved in the "Barletta" business group, it is now owned by a receiver which is in charge of selling it according to the GRP uses of public services.

Accordingly, the area is subject to a possible eviction due to the interest recently showed by a potential buyer. During the year, on the area has been construct a multi-factory with shared work spaces that became a reference social and cultural centre for the district.

7.5.2 Local Context

Officine Zero is located in Rome, in the so-called "Casal Bertone – Portonaccio" district which is part of the IV Municipality, considered by the citizens as the worst in Rome in terms of quality of life (Agenzia Roma Capitale 2017). Despite this perception, the "Casal Bertone-Portonaccio" district is facing years of profound urban changes (Fig. 7.8).

The district is characterised by the presence of the Tiburtina train-station, the second one of Rome in terms of number of connections and passengers from 2011 (La Repubblica 28/11/2011).

Moreover, the district is currently the subject of a huge real estate operation promoted by BNL–BNP Paribas Real estate bank group, which opened its new Italian headquarter (Dinamo Press 19/12/2017). In order to do that, the BNL–BNP Paribas Real estate bank group bought 75,000 m^2 of land near the Tiburtina train-station, changing the image of the district.

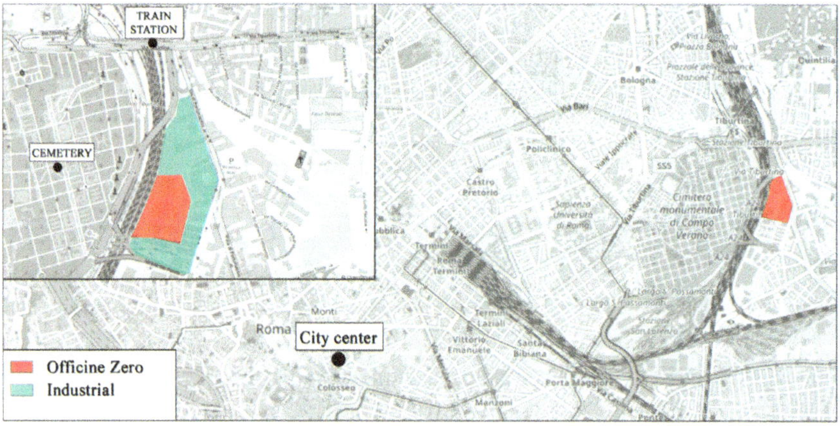

Fig. 7.8 The "Casal Bertone-Portonaccio" district

7.5.3 *Architectural Characters*

The OZ area is 29,000 m^2 of land and shows 22 different buildings, for a total of 9580 m^2 of covered spaces characterized by various dimensions and construction materials, depending on the use made across the years and the period of building. According to Colasanti (2013) and in line with the perception of the citizens, the 22 buildings are devoid of particular historical and cultural value.

It has wide external spaces, over 20,000 m^2, occupied by asphalt, green and trees, with connection to the tracks leading to Tiburtina train station.

All buildings are in poor condition, with untested facilities, but are equipped with basic urbanizations.

Among the 22 buildings present on the area, the main are (Fig. 7.9):

- A shed of 2451 m^2 characterized by a mezzanine that used to be for offices, consisting of iron trusses, reinforced brick floors, prefabricated partitions and cement floors, equipped with absent or untested plants;
- Eight warehouses with variable dimension from 800 to 100 m^2 showing the same characters of the shed;
- A keepers' house of three floors, brick and masonry, about 500 m^2, in poor condition;
- 5 masonry warehouses, in poor condition, surfaces between 10 and 100 m^2.

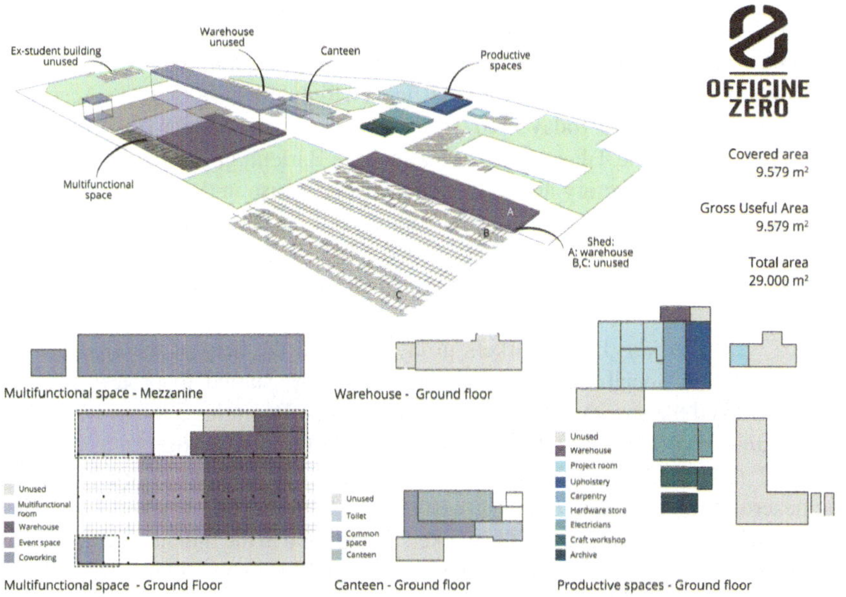

Fig. 7.9 The OZ scheme

According to the current occupants (personal interview 14/12/2018), the buildings need a profound restoration, roofs and plants above all, until today maintained by them, limiting their intervention to the used spaces.

7.5.4 *Origins and Stakeholders*

The Oz Officine Zero initiative born during the illegal occupation by former workers of RSI with the help of the component of Strike, a social centre located nearby. The first aim of Oz Officine Zero was to protest in order to grant a job to the workers. Only in a second time, the mission of the initiative changed to become a multi-factory.

Oz, as it is today, was in fact shaped by a group of about 30 people, living and working in the district, that entered the buildings after the illegal occupation started by the RSI workers.

The objective of these activists was first of all developing a political project regarding job and social conditions (personal interview 14/12/2018, ozofficinezero.org).

7.5.4.1 Pills of History

1930–2000: The owner of the area was "Compagnie Internationale des Wagons Lits" and its main activity was to provide night trains equipment;

2000: the property of the area and the business passed to the Rail Services International Italy S.P.A (RSI), which main activity was to maintain trains;

2008 (10 December): due to the failure of RSI, the area was acquired by the Barletta Group, whose main activity was logistics (La Repubblica 4/06/2013). According to the local press (Roma Today 5/07/2012) and to OZ (ozofficinezero.org), due to the projects regarding the Tiburtina trains station, the Barletta group started to dismiss the functions of the factory and requested a change of use in the GRP for the interested area, from industrial to residential/commercial uses (regional question No. 852 of 13/04/2012);

2011: the activities were interrupted and a lay off fund for the workers was activated;

2012 (20 February): the workers in lay off, with the help of the students and activists of the Social Centre Strike located nearby, started to protest and illegally occupied the spaces of the factory with the aim of granting a job to all workers (ozofficinezero.org; Martini 2015, Azzelini and Castronovo 2016);

2012 (24 July): The factory was entrusted to a bankruptcy trustee (ozofficinezero.org). According to the activists of OZ Officine Zero (personal interview 14/12/2018) only a small part of the workers found a job in the Italian railway system, while the others decided to continue the occupation. Meanwhile some activists of the district, interested in the topic of job rights and new models of economic development, joined the occupation carrying the project of constructing a reuse and recycle centre in the spaces of the factory. The aim of this free initiative was to grant a job to the workers as well as to other interested people;

2013 (1 June): OZ opened illegally occupying the area and following the southern American model called "Fabricas Recuperadas". This kind of model provides to:

- reactivate the factory, in this case with a different production, reusing waste materials;
- develop the activities with the direct responsibility of the workers and people using the spaces;
- communicate and discuss, meanwhile, topics such as work rights and environmental sustainability, developing an actual political project.

The organization model is horizontal: general assembly for the core decisions and mandatory assemblies for each part of the community (hozofficinezero.org; Martini 2015; Azzelini and Castronovo 2016).

In that period new workers and occupants arrived from the neighbourhood. They joined the OZ spaces after a selection based on the proposed project topics and objectives. The spaces in 2013 were mainly used as coworking spaces for professionals and artisans (personal interview 14/12/2018).

2014–2015: the first auction for the purchase of the area was launched by the bankruptcy trustee. The starting price was 2.1 million € (Colasanti 2013), but no one was interested. Nonetheless a public mobilisation started defending the OZ Officine Zero project (change.org 2015; Dinamo press 20/11/2015; Il Manifesto 24/11/2015).

During the years the project changed. The idea of creating a reuse centre failed due to the needs of the occupant workers: they worked as normal autonomous professional, each one with his own interests but not with the only aim of recycling in a common framework. In this way, Officine Zero gradually became a multi-factory where professionals, students and activists, discussed and projected a new model of working and living, developing projects and events for the whole district and the city.

In order to facilitate and to be able to participate in "legal" calls and projects, they created a non-profit association called Zero Off, functional to the development of the informal network of Officine Zero (personal interview 14/12/2018).

April 2017: The BNL–BNP Paribas Group declared its interest in buying the area;

The group had already invested in the district, starting a real estate operation in 2011, with the construction of its head quarter next to the Tiburtina trans station.

In order to avoid the possible purchase and therefore the closing of Officine Zero, the occupants requested to the Municipality of Rome the declaration of the public utility of the project and of the interested spaces. The request was rejected due to the private nature of the property of the interested area (Roma Today 10/04/2017; personal interview 14/12/2018).

2017 (1 June): The public auction was opened for the sale of the lot and the BNL group officially expressed its interest (Roma today 6/02/2018). According to the activists (personal interview 14/12/2018) from that moment a bargaining process started between Officine Zero, BNL Group and the institutions. In fact, given the initial rejection of the Municipality of Rome, the activists involved the Lazio Region which assumed the role of intermediary.

June 2018: The Municipality of Rome officially declared the spaces of OZ under eviction. This step was widely reported by the local and national press, among others La Stampa (27/06/2018), rAdnkronos (26/06/2018) and Il Fatto quotidiano (8/07/2018);

However, the dialogue between the parts continued. The Lazio Region and the Municipality of Rome in fact would like to reach a peaceful agreement between the property and the occupants, before the development of the BNL Group project (activists, personal interview 14/12/2018);

2018 (November–December): According to the activists involved in the bargaining process (personal interview 14/12/2018), the BNL–BNP Paribas Group is considering the possibility of granting the future development of OZ assuring a new space eligible to continue its social and working activities.

7.5.5 Value Proposition

OZ currently stands as a multi-factory: a space dedicated to shared work, managed directly by the workers/professionals. In particular, according to the 2017 report of activities (ozofficinezero.org), the main activities are:

- productive spaces, which hosts about 50 workers including: carpenters, blacksmiths, restorers, up cyclists, upholsterers, designers, a journalistic editorial staff, a sustainable tourism association, an association of social designers, a studio of photographers and printers, a bicycle shop, a collective of architects;

- a coworking space for freelancers;
- a canteen;
- a multifunctional space for events;
- open spaces used for sports and cultural activities;
- calendar of musical events, exhibitions and artistic festivals;
- projects of social nature on the district and the city.

7.6 Ex Fadda, San Vito Dei Normanni

7.6.1 In Brief

Ex Fadda (exfadda.it) is a cultural centre and social innovation laboratory located in San Vito Dei Normanni, Brindisi (Italy) that used to be an oenological factory built by a local noble family called Dentice di Frasso. During the years it was donated to the Municipality of San Vito Dei Normanni and it was recovered thanks to the "Bollenti Spiriti" program of the Puglia Region. It is currently managed by a local

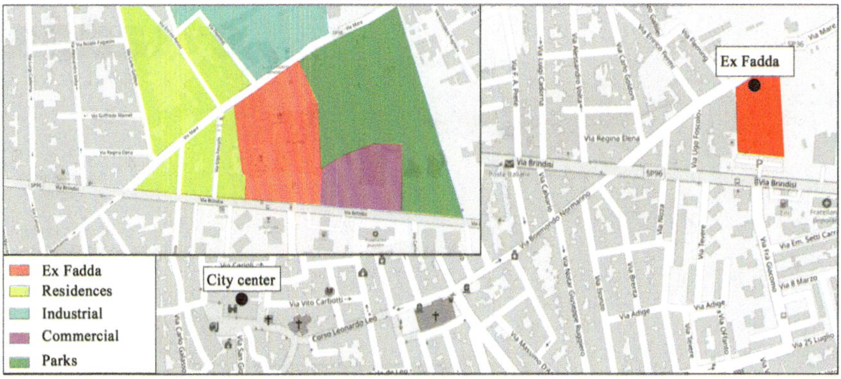

Fig. 7.10 The "Ex Fadda" area in San Vito dei Normanni

communication company, the "Sandei Srl", and the social cooperative "Qualcosa di diverso".

Ex Fadda hosts cultural activities, events, music, dance, handicraft schools, a bar, a restaurant and various local associations characterised by youth engagement and promotion. It is moreover linked to the so called "XFarm", an olive and wine farm seized from crime organizations.

7.6.2 Local Context

San Vito dei Normanni is a city of 19,000 inhabitants (ISTAT 2018) located in the well-known Salento hinterland in the Puglia region. It is characterized by a predominantly agricultural and commercial economy and a demographic structure strongly marked by youth emigration. From the '50s to 2000 the town hosted a United States air force base, then dismantled and now used by UNHRD (Brindisi Oggi 17/06/2014).

Ex Fadda is located on the main road entering the city next to the historical centre, in the nearby is possible to find some commercial activities and restaurants (Fig. 7.10).

The area in which Ex Fadda stands has public and services destination coherent with the GRP of San Vito dei Normanni.

7.6.3 Architectural Characters

The Dentice di Frasso oenological factory was built at the beginning of the 20th century by the homonymous local noble family.

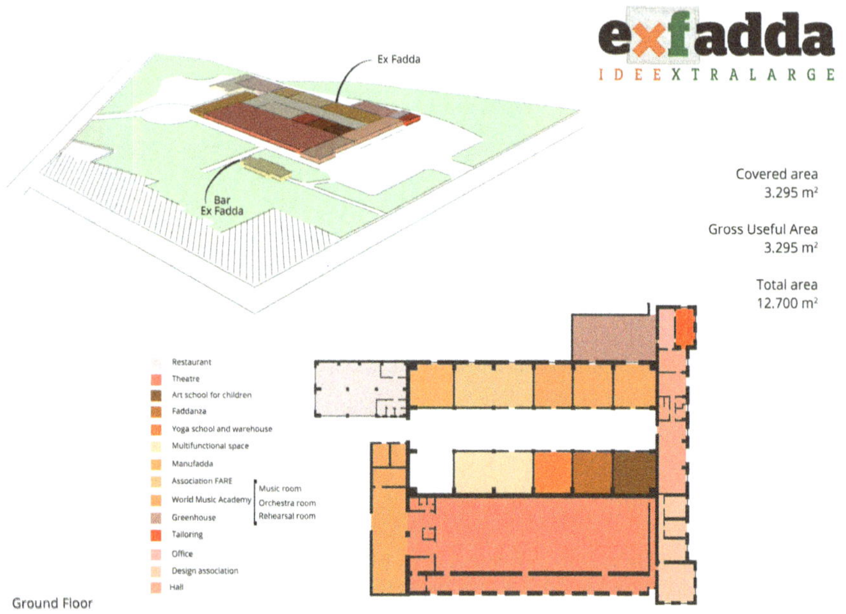

Fig. 7.11 The Ex Fadda scheme

The buildings are located in an area of 12,700 m^2 of land, of which 3300 m^2 of covered surface are divided in two main parts (extramagazine.eu): the masonry (about 2000 m^2) has 18 vaults "a sharp corner"; the second (about 1300 m^2) has trusses wooden roofing. The first has been divided in smaller spaces with drywall, except a central vaulted multifunctional space (Fig. 7.11).

The external area hosts a small bar, in the former guard house, is completely fenced, and is now used as garden, car park and event area.

7.6.4 *Origins and Stakeholders*

The Municipality of San Vito dei Normanni applied in 2010 the "Laboratori Urbani" programme promoted by the Puglia Region in the framework of "Bollenti Spiriti" programme dedicated to youth development. This program was created and funded by the Region in order to reactivate abandoned public buildings all over the region with youth engagement and social innovation projects. (bollentispiriti.regione.puglia.it; Tricarico 2014).

With the Municipality, a group of 6 local organizations participated in the "Bollenti Spiriti" call.

The amount of funding coming from the call for the activation of the Ex Fadda project was 50,000 €. The great part of the works needed for the project were done by voluntaries and activists during the self-construction workshops promoted from 2011 to 2012 (personal interview 26/01/2019; bollentispiriti.regione.puglia.it; lascuoladibollentispiriti.it).

The initial goals were:

- the bottom up regeneration of the space;
- informal learning, activating the network of associations and citizens present with the aim of building common skills, linked both to the design of institutional partnerships and to new practices of use of space;
- the construction of a "horizontal" community that sees a mix of roles between users and operators of the same activities, proposing continuous initiatives to open up the spaces to the local community (exfadda.it).

7.6.5 History Pills

Early 1900s: Construction of a wine and food factory commissioned by the Dentice di Frasso family;

End of the '50s: The factory has been donated to the Municipality of San Vito dei Normanni with the bound of using it for social and public aims. From that period the buildings were used as warehouse, public housing and then abandoned in a bad state of maintenance (personal interview 26/01/2019);

2008–2010: The Municipality restored the building with an investment of 350,000 € (ilgiornaledellefondazioni.it 15/11/2015);

The municipality of San Vito dei Normanni inserted the structure in the project "Bollenti spiriti", won with 5 years' trust, for the management part, by the "Sandei Srl" (leader) communication company together with other 5 local organizations (Tricarico 2014);

2010: Opened the Ex Fadda Café in the former guard house;

2011: The 70% of the spaces were still unused. The "Sandei Srl" decided to hire Roberto Covolo (engaged in the central management of Bollenti Spiriti programme for the Puglia Region) as project manager, in order to reactivate the development and innovation path of Ex Fadda (ilikepuglia.it 23/04/2014);

2012 (20 September–20 October): The Ex Fadda self-construction laboratories were organized in order to reactivate and restore all the spaces with the participation of local population;

2012: the recovery of the spaces was completed and Ex Fadda began to host associations and initiatives led by local people (Giannini 2017; Notarnicola 2016; Corriere della Sera 16/6/2014);

Morever opened XFood, a restaurant born thanks to the Puglia Region funds dedicated to integrated innovative projects for the social inclusion of disadvantaged people;

2014: the social cooperative "Qualcosa di diverso" was born for the management of the activities inside Ex Fadda (President Roberto Covolo), while the buildings were still managed by the "Sandei Srl" (Giannini 2017; Notarnicola 2016);

2015: the call for the management was renewed by the Municipality of San Vito dei Normanni and again entrusted to "Sandei Srl" until 2021 (Giannini 2017; Notarnicola 2016);

2016/2017: "Qualcosa di diverso" cooperative opened the XFarm, a farm producing olive oil and wine, after the assigning of land seized to organized crime (personal interview 26/01/2019).

7.6.6 Value Proposition

The main activities currently hosted and developed by Ex Fadda are (exfadda.it) are:

- *XFarm*: a farm producing wine and olive oil;
- *XFood*: a restaurant;
- *Ex Fadda Caffè*: a bar;
- *World music academy*: a music training and production centre;
- *Manufadda*: a community craft project that connects the design skills of young designers and the manual skills of ladies from the artisan territory for hobbies;
- *Noninlinea*: a collective of photographers and videomakers engaged in a research on visual storytelling;
- *Faddance*: a school of classical and contemporary dance promoted by a young dancer of the territory;
- *Fare*: a school for young artisans;
- *Lamusicadentro*: a music project for children from 0 to 3 years and their families;
- Common garden: a garden designed, built and managed by citizens;
- *Santu Vitu mia*: a program of public participation and consultation on issues of urban regeneration at the local level;
- *La Chimera*: a school of contemporary art for children;
- *Madeinterra*: an innovative start-up that deals with green building and the use of natural materials in architecture;
- *Da Zero a Uno*: an orientation and support desk to turn ideas into projects and start to realize them;
- *Da grande sarò un teatro*: creation of a production and performance centre for performing arts.

7.7 Farm Cultural Park

7.7.1 *In Brief*

The Farm Cultural Park, also called simply Farm (farmculturalpark.com) is an independent and private cultural centre born in 2010 in the small city of Favara in Sicily. The aim of the Farm Cultural Park is to provide an answer to the deterioration of the historic centre of Favara and of the quality of life for the population living there. In order to do that, the Farm aims to redevelop Favara through art and culture, improving the quality of life. The Farm is composed of a series of connected buildings located in the historical city centre, called "Sette Cortili" or "Cortile Bentivegna". The buildings are owned by the association and the people composing the Farm and host exhibition spaces, residences, artistic installations and a school of architecture for children.

7.7.2 *Local Context*

Favara is an Italian town in the south of Sicily of 30,000 inhabitants (ISTAT 2018), located 8 km from the UNESCO site Valle dei Templi and 10 km from Agrigento. It is related to intense socio-economic interaction with this city, with which it forms a conurbation (Fig. 7.12).

The Farm is located in the historical city centre of Favara, next to the main square (Piazza Cavour) populated by numerous restaurants and touristic activities. The Farm involves the entire "Sette Cortili" district with an urban morphology with Arabic roots.

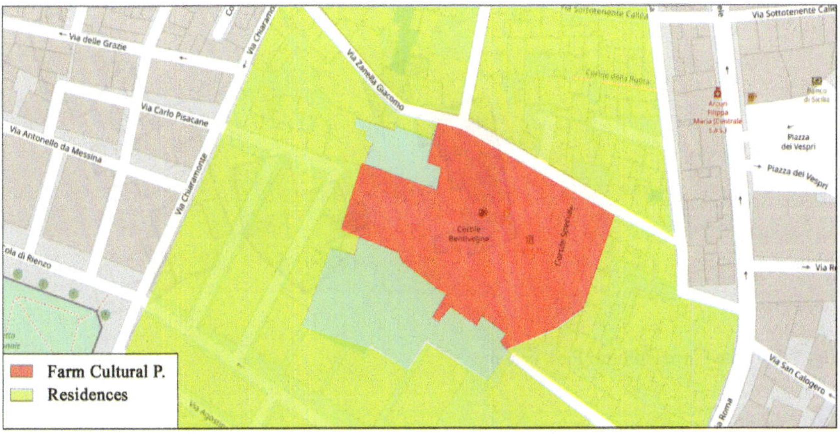

Fig. 7.12 The Favara town

Starting from the 90s, the district has been abandoned and therefore it is currently socially and structurally degraded due to the problems deriving from the building interventions without urban rules and plans (Di Carlentini and Liotta 2016; Occhipinti 2017; Faraci 2017).

7.7.3 Architectural Characters

The district hosting the Farm is about 18,000 m^2. The occupied area consists of seven connected courtyards forming the Cortile Bentivegna and hosting small historic palaces and gardens. The covered area of the interested renovated buildings is about 1750 m^2, while 2550 m^2 have been acquired but they are not restored yet. (farmculturalpark.com).

The Farm (Fig. 7.13) consists now of about ten structures: a small building with the gallery and a shop were the first two structures opened in June 2010. Every year more spaces are added. Each property has its intended use and its precise theme. Although the spaces change appearance continuously, thanks to exhibitions and events (Di Carlentini and Liotta 2016; Occhipinti 2017; Faraci 2017).

Fig. 7.13 The Farm Cultural Park scheme

7.7.4 *Origins and Stakeholders*

The Farm cultural park is a private initiative, started by two private citizens: Andrea Bartoli and Florinda Saieva. They are a married couple from Favara and the nearby, (their jobs: Notary him and Attorney of the Sacred Rota her), who chose not to emigrate to deal with the degradation of their city and grow their two daughters. In order to buy and restore the first nucleolus of interested buildings they invested around 500,000 €. (Personal interview 16–17 November 2018; Sicilian Post 20/09/2018)

7.7.4.1 Pills of History

Since the 60s: due to economic and commercial dynamics, the historic centre of Favara was subject to depopulation and degradation of structures: the historic centre with seven courtyards becomes a ghetto-district subject to crime and social hardship. Moreover, due to the lack of a GRP, the city has developed in an uneven way from an urban point of view (Di Carlentini and Liotta 2016; Occhipinti 2017, Faraci 2017);

2008–2009: Andrea Bartoli and Florinda Saieva (personal interview 16–17/11/2018) was living between Paris (France) and Favara: Andrea worked as a Notary in Sicily, while Florinda and her daughters were settled in Paris. This situation was difficult for them, cause they wanted to raise their children together, in a high quality environment from a social and cultural point of view. They decided to settle in Favara creating there the stimulus context they wanted for their children and working for a better quality of life for the whole population of Favara. They bought the buildings in Cortile Bentivegna using their money (the expense was about 500,000 €) and they projected the Farm Cultural Park, planning to open it in 2012;

2010 (23 January): A building collapsed in the historic centre due to poor conservation causing the death of two young girls (La Repubblica 23/01/2010). This dramatic fact brought the PA of Favara to start demolishing the historic centre;

2010 (March): Due to the dramatic situation, Andrea and Florinda decided to speed up the work (personal interview 16–17/11/2018). They started the restructuring of two buildings to give a signal to the population and to stop the process of demolition of the historic centre;

2010 (28 June): The Farm Cultural Park opened with the XL building: a building dedicated to exhibition space and art shop, but also a related cultural and artistic events programme. Accordingly, the Farm Cultural Park started growing with new spaces, bought or rented by Andrea, Florinda and their association, exhibitions and the opening of a school of architecture for children. All the activities planned had an international scale. Farm Cultural Park hosts well-known artists and architects, attracting new public and tourism, but also universities and researchers. The Farm started attracting the attention of local, national and international press (among others The Guardian (9/03/2012), Artribune (24/06/2011), La Repubblica 21/02/2015, Il sole 24 ore (7/02/2016), ANSA (22/06/2017), Lonely Planet (2017);

2014(28 June): The community cooperative "Farmidabile" was founded to bring together citizens and private operators linked to Farm Cultural Park;

"Formidable" was the result of a number of activities funded thanks to the tourists and the public attracted by Farm Cultural Park as restaurants, B&B, hotels, commercial activities opened in the city centre, around the Cortile Bentivegna. This impact was roughly quantified, by Andrea Bartoli (personal interview 16–17/11/2018, in around 20 millions €. The most visible result of this process is the main square of the historic centre, Cavour square, where since 2010 (Farm opening) flourished new bars and restaurants;

2017 (24 July): The Municipality of Favara, which in the past years was indifferent to the project, namely the local police, issued an eviction order for illegal occupation of public land. With this order it condemned Farm Cultural Park to pay a fine and to take apart an artistic installation located in a courtyard in Cortile Bentivegna. These happenings determined a crisis between the Farm and the municipality, but also the mobilitation of citizens, local and national press in defense of Farm Cultural Park (Artribune 31/07/2017, change.org 2017, Il Sicilia 29/07/2017, Il Fatto Quotidiano 2/08/2017, AgCult 31/07/2017);

2017 (18 August): After the mobilization, at national and international level, of the public, the media, and the artistic community around Farm, the municipality of Favara issued the resolution of declaration of the public utility of Farm Cultural Park project, withdrawing the order and approving a resolution to cooperate in the development of the historic centre (Comune di Favara 18/08/2017, Sicilia On Press 19/08/2017);

2018: After eight years of development in Favara, Andrea and Florinda launched SPAB "Società per azioni buone", with the aim of enlarging and replicating the experience of Farm Cultural Park, through the creation of a joint stock company, in which the funds raised among simple citizens can be used to activate investments of social and cultural nature, in order to increase the quality of life in Favara first of all, but also elsewhere (Il Giornale delle fondazioni 15/10/2018).

7.7.5 Value Proposition

According to Andrea Bartoli (personal interview 16–17/11/2018) the main objective of Farm Cultural Park is still to facilitate the development of Favara and its historical centre through art and culture, and it does so through the following spaces and activities:

- Calendar of cultural events of various kinds and exhibitions;
- *Farm XL*, main building, exhibition centre and base for cultural projects;
- *SOU*, School of Architecture for children;
- Artist residences;
- *Holy Cow*, a coworking space;
- *Nzemulla*, shared kitchen and photographic space;

- *Barbecue Garden*, garden used as an event space;
- *Stay Hungry*, *Stay Foolish*, motivational public toilet;
- *Palazzo Giglia*, under renovation for events space, coworking and hostel.

To these components, between 2017 and 2018, was added the project of "Società per azioni buone," a joint-stock company, with the aim of starting up urban regeneration processes, based on the Farm model, through popular ownership (Sicilian Post 20/09/2018).

References

Agnoli, E. (2014). Spazio Grisù: recupero creativo e riattivazione produttiva a Ferrara, Imprese & città, N. 4/2014, Camera di commercio di Milano.

Agnoli, E. (2015). Fiamma creativa: da ex caserma a spazio creativo nel centro di Ferrara. In: I quaderni della Fondazione Brodolini, n. 51/2015, Fondazione Giacomo Brodolini.

Azzelini, D., & Castronovo, A. (2016). Fabbriche recuperate e nuova istituzionalità mutualistica. In Sindacalismo sociale. Lotte e invenzioni istituzionali nella crisi europea, DeriveApprodi.

Balestra, A., & Ferrero M. (2016). Area OSI OVEST-NORD: Toolbox Coworking, in "Postfordismo e Trasformazione Urbana. Casi di recupero dei vuoti industriali e indicazioni per le politiche nel territorio torinese", Ires Piemonte.

Colasanti, D. (2013), Perizia Costa Sistemi Ferroviari, Tribunale di Lecco, N.31/2013.

Di Carlentini, M., & Liotta, S. J. (2016). Farm Cultural Park, Planur-e, n8/2016.

Faraci, G. (2017). Farm Cultural Park: an experience of social innovation in the recovery of the historical centre of Favara, Procedia Environmental Sciences, n37 2017, pp. 676–688. Elsevier.

Giannini, F. (2017). Il Laboratorio Urbano Ex Fadda a San Vito dei Normanni. Indagine etnografica e prospettive di applicazione dell'antropologia in un community hub, Graduation thesis, University of Bologna.

Guzzon, A. (2012). "Spazio Grisù": una fabbrica della creatività e della cultura, La Pianura n. 3, 2012, Camera di commercio di Ferrara.

Mangialardo, A., & Micelli, E. (2017). Simulation models to evaluate the value creation of the grass-roots participation in the enhancement of public real-estate assets with evidence from Italy. *Buildings, 7*(4), 100.

Martini, L. (2015). Il caso degli spazi workers-control: Officine Zero, tra autoproduzione e territorio. Atti della XVIII Conferenza nazionale SIU, Italia '45-'45, Venezia 11–13 giugno 2015, Planum Publisher, Roma-Milano 2015.

Monti, M. (2010). Estratto di perizia di stima, Province of Ferrara.

Municipality of Favara. (2017). Deliberazione della Giunta comunale del 18/08/2017.

Municipality of Ferrara. (2016). Assegnazione in sub comodato al consorzio Factory Grisù di porzione dell' immobile sito in via Poledrelli n. 21, Prot. Gen. n. 67468/2016.

Municipality of Terni. (2016). Terni Smart City. Agenda Urbana 2014–2020.

Notarnicola, M. (2016). Il caso ExFadda: un'esperienza di sviluppo delle capabilities, Graduation thesis, University of Bologna.

Occhipinti, F. (2017). Farm Cultural Park come laboratorio di rigenerazione territoriale, Agathon, 1/2017.

Sposito, C. (2012). Sul recupero delle aree industriali dismesse. Tecnologie materiali impianti ecosostenibili e innovativi, Maggioli Editore.

Tarquini, A. (2002). Terni i Programmi Urbani Complessi, Municipality of Terni.

Tricarico, L. (2014). Imprese di Comunità nelle Politiche di Rigenerazione Urbana: Definire ed Inquadrare il Contesto Italiano, Euricse Working Papers.

Sitography

Adnkronos. (26/06/2018). Sgombero imminente: l'appello di Officine Zero. adnkronos.com/fatti/cronaca/2018/06/28/sgombero-imminente-appello-officine-zero_I6Q55Fwh0ZizNu9pSvO6cJ.html?refresh_ce. Accessed July, 2018.
AgCult. (31/07/2017). Farm Cultural Park Favara, mercoledì incontro alla Camera coi responsabili Cultura del Pd. agcult.it/2017/07/31/farm-cultural-park-favara-mercoledi-incontro-alla-camera-coi-responsabili-cultura-del-pd/. Accessed July 20, 2018.
Agenzia Roma Capitale. (2017). Indagine sulla qualità della vita e dei servizi pubblici locali a Roma (X edizione). agenzia.roma.it/documenti/monitoraggi/530.pdf.
ANSA. (22/06/2017). Farm Cultural Park, così rinasce paese dell'agrigentino. ansa.it/sicilia/notizie/2017/06/22/ansa-farm-cultural-park-cosi-rinasce-paese-dellagrigentino_3095bd5a-6597-466b-96fe-ca86fe345891.html. Accessed July 19, 2018.
Archdaily.com. (14/10/2010). Toolbox/Caterina Tiazzoldi, archdaily.com/81630/toolbox-caterina-tiazzoldi. Accessed June 20, 2018.
Artribune. (24/06/2011). Irripetibili. È tempo di arte-e-musica a Favara. artribune.com/arti-performative/musica/2011/06/irripetibili-e-tempo-di-arte-e-musica-a-favara/. Accessed July 19, 2018.
Artribune. (31/07/2017). "Quelle installazioni sono abusive". La burocrazia s'accanisce sul Farm Cultural Park di Favara. artribune.com/arti-visive/arte-contemporanea/2017/07/quelle-installazioni-sono-abusive-la-burocrazia-saccanisce-sul-farm-cultural-park-di-favara. Accessed July 20, 2018.
Bollenti Spiriti. (2010). http://bollentispiriti.regione.puglia.it/index.php?option=com_jumi&view=application&fileid=1&prog=312&Itemid=448. Accessed July 10, 2018.
Change.org. (2015). OZ Officine Zero - Non si chiude con l'asta. change.org/p/roma-enti-locali-oz-officine-zero-non-si-chiude-con-l-asta. Accessed July 13, 2018.
Change.org. (2017). Noi siamo Farm Cultural Park. change.org/p/noi-siamo-farm-cultural-park. Accessed July 20, 2018.
Civita.it. (28/03/2009). CAOS, Centro per le Arti Opificio Siri inaugura con "Map Games: Dynamics of Change". civita.it/Sala-stampa/CAOS-Centro-per-le-Arti-Opificio-Siri-inaugura-con-Map-Games-Dynamics-of-Change. Accessed June 25, 2018.
Corriere della Sera. (16/06/2014). ExFadda, l'officina che crea il futuro. nuvola.corriere.it/2014/06/16/exfadda-il-laboratorio-che-crea-il-futuro/. Accessed July 10, 2018.
Dinamo Press. (19/12/2017). BNP Real Estate a Tiburtina: la città a misura di banca. dinamopress.it/news/bnp-real-estate-tiburtina-la-citta-misura-banca/. Accessed July 13, 2018.
Dinamo Press. (20/11/2015). Il progetto di Officine Zero è a rischio! dinamopress.it/news/il-progetto-di-oz-officine-zero-e-a-rischio/. Accessed July 13, 2018.
Divisare.herokuapp.com. (2010). Toolbox coworking, divisare.herokuapp.com/projects/132323-caterina-tiazzoldi-helene-cany-toolbox-coworking-torino. Accessed June 20, 2018.
Estense.com. (7/11/2017). Il quartiere 'Giardino' è sempre più periferia degradata. estense.com/?p = 651509. Accessed Sept 12, 2018.
exfadda.it. Accessed July 10, 2018.
Extra Magazine. (28/02/2014). ExFadda/Nuove storie per antichi luoghi, extramagazine.eu/it/blog/6-cultura/2063-exfaddanuove-storie-per-antichi-luoghi.html. Accessed July 10, 2018.
factorygrisu.it/, Bandi. Accessed September 13, 2018.
ozofficinezero.org/. Accessed July 13, 2018; Feb 25, 2019.
caos.museum/. Accessed June 25, 2018.
farmculturalpark.com/. Accessed July 19, 2018.
tiazzoldi.com. (2010). Toolbox project. Accessed June 20, 2018.

I like Puglia.it. (23/04/2014). Ripartire da zero, la storia di Roberto e di 'ExFadda', ilikepuglia.it/notizie/sviluppo/bari/23/04/2014/ripartire-da-zero-la-storia-di-roberto-e-di-exfadda.html. Accessed July 10, 2018.

Il Fatto Quotidiano. (2/08/2017). Agrigento: cara sindaca di Favara ci ripensi, non è abusivismo ma rigenerazione. ilfattoquotidiano.it/2017/08/02/agrigento-cara-sindaca-di-favara-ci-ripensi-non-e-abusivismo-ma-rigenerazione/3770934/. Accessed July 20, 2018.

Il Fatto quotidiano. (8/07/2018). Roma, dal fallimento RSI alla cessione dell'area. I lavoratori delle 'Officine zero' a rischio sgombero: "Si faccia chiarezza". ilfattoquotidiano.it/2018/07/08/roma-dal-fallimento-rsi-alla-cessione-dellarea-i-lavoratori-delle-officine-zero-a-rischio-sgombero-si-faccia-chiarezza/4458855/. Accessed July 13, 2018.

Il Giornale delle fondazioni. (15/10/2018). SPAB – SOCIETÀ PER AZIONI BUONE: un nuovo progetto di Farm Cultural Park. http://ilgiornaledellefondazioni.com/content/spab-%E2%80%93-societ%C3%A0-azioni-buone-un-nuovo-progetto-di-farm-cultural-park. Accessed November 10, 2018.

Il Manifesto. (24/11/2015). Non bruciate il sogno delle Officine Zero. ilmanifesto.it/storia/non-bruciate-il-sogno-delle-officine-zero/. Accessed July 13, 2018.

Il Sicilia. (29/07/2017). Farm Cultural Park, la strana ordinanza del comune di Favara che considera l'arte un abuso. ilsicilia.it/farm-cultural-park-la-strana-ordinanza-del-comune-di-favara-che-considera-larte-un-abuso/. Accessed July 20, 2018.

Il sole 24 ore. (7/02/2016). Storie di un notaio «pazzo» e dei ragazzi di Favara. ilsole24ore.com/art/cultura/2016-02-07/storie-un-notaio-pazzo-e-ragazzi-favara-081518.shtml?uuid=ACktGUPC. Accessed July 19, 2018.

ilgiornaledellefondazioni.it. (15/11/2015). Ex Fadda: innovazione culturale formato comunità. http://ilgiornaledellefondazioni.com/content/exfadda-innovazione-culturale-formato-comunit%C3%A0. Accessed July 10, 2018.

ISTAT. (2018). Popolazione residente comunale per sesso anno di nascita e stato civile. http://dati.istat.it. Accessed February 20, 2019.

La Nazione.it. (8/10/2014). Terni, il bando per i servizi museali lo vincono ancora Civita Cultura e Indisciplinarte. lanazione.it/umbria/cronaca/terni-caos-gara-bando-servizi-museali-siri-1.285379. Accessed June 25, 2018.

La Repubblica. (23/01/2010). Favara,crolla palazzina di tre piani. Due bambine morte, salvo il fratello. repubblica.it/cronaca/2010/01/23/news/favara_crollo-2047962/. Accessed July 19, 2018.

La Repubblica. (28/11/2011). Tiburtina, via alla nuova stazione; il primo hub per l'Alta velocità. roma.repubblica.it/cronaca/2011/11/28/news/tiburtina_via_alla_nuova_stazione_annunciati_presidi_dei_no_tav-25710278/. Accessed July 13, 2018.

La Repubblica. (4/06/2013). Nascono le Officine zero dalle ceneri di Wagon Lits. ricerca.repubblica.it/repubblica/archivio/repubblica/2013/06/04/nascono-le-officine-zero-dalle-ceneri-di.html. Accessed July 13, 2018.

La Repubblica. 21/02/2015. Farm Cultural Park, l'arte contemporanea che da Favara conquista il mondo. palermo.blogautore.repubblica.it/2015/02/21/farm-cultural-park-larte-contemporanea-che-da-favara-conquista-il-mondo/. Accessed July 19, 2018.

La scuola di Bollenti Spiriti. (2011). Project work Ex Fadda. http://lascuoladibollentispiriti.it/project-work/exfadda. Accessed July 10, 2018.

La Stampa. (27/06/2018). Ordinanza di sgombero per "Of Officine Zero" l'ex fabbrica di Portonaccio riconvertita in multifactory. lastampa.it/2018/06/27/roma/ordinanza-di-sgombero-per-of-officine-zero-lex-fabbrica-di-portonaccio-riconvertita-in-multifactory-0czPwZF1z36AYBXnip0NZL/pagina.html. Accessed July 13, 2018.

Lonely Planet. (2017). Farm Cultural Park. lonelyplanet.com/italy/favara/attractions/farm-cultural-park/a/poi-sig/1401121/1340858. Accessed July 19, 2018.

Officine Zero. (2017). Dossier on the activities of 2017. drive.google.com/file/d/1Ykjr346GSj5tScPd3WHmOC3RNKgNrt-9/view. Accessed July 13, 2018.

Presstletter.com. (25/03/2011). Progetto finalista Premio Fondazione Renzo Piano ad un giovane talento, presstletter.com/2011/03/progetto-finalista-premio-fondazione-renzo-piano-ad-un-giovane-talento-5/. Accessed June 20, 2018.
Professionearchitetto.it. (11/5/2011). Campus IED: un progetto firmato Mario Cucinella, professionearchitetto.it/news/notizie/13290/Campus-IED-un-progetto-firmato-Mario-Cucinella. Accessed June 20, 2018.
Ricerca.repubblica.it. (6/12/2016). Benvenuti a "Toolbox" il regno del coworking "Qui il lavoro del futuro", ricerca.repubblica.it/repubblica/archivio/repubblica/2016/12/06/benvenuti-a-toolbox-il-regno-del-coworking-qui-il-lavoro-del-futuroTorino15.html. Accessed June 20, 2018.
Roma Today. (10/04/2017). A Portonaccio le Officine Zero, da fabbrica a 'multifactory': "Diventino di pubblica utilità". romatoday.it/politica/officine-zero-via-partini-portonaccio.html. Accessed July 13, 2018.
Roma Today. (5/07/2012). Tiburtino: fabbrica occupata da 5 mesi "per difenderci dalla speculazione edilizia". tiburtino.romatoday.it/occupazione-fabbrica-wagon-lits.html. Accessed July 13, 2018.
Roma Today (6/02/2018). L'area di Officine Zero fa gola a Bnl: la banca allarga le sue mire su Casal Bertone. romatoday.it/politica/officine-zero-bnl-bnp-paribas.html. Accessed July 13, 2018.
Sicilia On Press. (19/08/2017). FARM ecco la convenzione che regolamenterà l'ingresso ai sette cortili. siciliaonpress.com/comune-di-favara-farm-ecco-la-convenzione-che-regolamentera-lingresso-ai-sette-cortili/. Accessed July 20, 2018.
Sicilian Post. (20/09/2018). Il caso Farm di Favara, utopia reale di Bartoli «Il segreto? Ho agito». sicilianpost.it/il-caso-farm-di-favara-utopia-reale-di-bartoli-il-segreto-ho-agito/. Accessed November 10, 2018.
spaziogrisu.it/, L'associazione. Accessed September 12, 2018.
Terni Today.it. (3/09/2018). Caos, si tratta per un taglio ai costi della gestione dei musei. ternitoday.it/politica/terni-caos-taglio-spese-gestione-musei.html. Accessed November 28, 2018.
The Guardian. (9/03/2012). Sicily: the art project that saved a town. theguardian.com/travel/2012/mar/09/sicily-favara-art-farm-cultural-park. Accessed July 19, 2018.
Torino guarda in alto. (2007). toguardainalto.altervista.org. Accessed June 20, 2018.
Umbria 24. (16/11/2013). Terni, «Città giardino futuro 'quartiere creativo'». Idee e dubbi per l'urbanistica del futuro, umbria24.it/attualita/terni-citta-giardino-futuro-quartiere-creativo-idee-e-dubbi-per-lurbanistica-del-futuro. Accessed June 25, 2018.
Umbria 24.it. (2/3/2017). Cultura a Terni, bene Carsulae mentre il Caos è un flop ma Trimarchi non lo dice. umbria24.it/cultura/cultura-terni-bene-carsulae-caos-un-flop-trimarchi-non-lo-dice. Accessed June 25, 2018.
Umbria24.it. (18/12/2013). Terni, Indisciplinarte e la gestione del Caos: «Tutt'altro che un carrozzone, ecco i dati». umbria24.it/attualita/terni-indisciplinarte-e-la-gestione-del-caos-tuttaltro-che-un-carrozzone-ecco-i-dati. Accessed June 25, 2018.
Umbria24.it. (24/05/2017). Terni, il Caos conta 21 mila visitatori ma pagano solo in 6 mila. umbria24.it/cultura/caos-terni-21-mila-visitatori-pagano-solo-6-mila-trimarchi-numeri. Accessed June 25, 2018.
Umbria24.it. (8/08/2014). Terni, Caos: un bando che provoca polemiche. Scontro in vista in consiglio comunale. umbria24.it/politica/terni-caos-un-bando-che-provoca-polemiche-scontro-in-vista-in-consiglio-comunale. Accessed June 25, 2.

Chapter 8
Governance, Economic Sustainability and Socio-spatial Relationships

Sara Bonini Baraldi and Carlo Salone

Abstract The chapter summarizes major findings emerging from the six case studies. Drawing on the analytical framework presented in chapter five, the chapter is organized in three sections. The first section—"Genesis, governance and organisation"—underlines the experimental drive and, consequently, the developmental nature of the projects analysed. The "business model and economic sustainability section" highlights how all initiatives are characterized by the stratification of multiple values, a kind of "value mix" which m responds to both strictly productive as well as social, and at times cultural, needs. The value mix becomes a fundamental component of the business model, clearly reflecting on the definition of the users of the initiatives and the type of activity performed, as well as the economies generated/consumed. The "socio-spatial relations" generated by the six cases are analysed in section three, focusing on both the relations with the 'local' and 'supra-local' networks, and with local public powers. Final remarks reflection the ownership of the assets in question, checking if and how this differentiates the cases in terms of governance, legal form and organisation on one hand, and mission, business model and sustainability on the other.

The experiences presented and discussed here were selected on the basis of the criteria explained in the methodological chapter, which allowed us to view the projects analytically. The cases were selected to offer a representative overview of the experiences of adaptive reuse and entrepreneurial self-organisation.

While the first and last dimensions highlighted in the analytical chart (architectural features and urban context; market revaluation of the asset) are examined elsewhere, here we will focus on the remaining four analytical dimensions (origin of the initiative; governance and organisation; business model and sustainability; socio-spatial

S. Bonini Baraldi (✉) · C. Salone
Department of Regional and Urban Studies and Planning (DIST), Politecnico di Torino and Università di Torino, Viale Mattioli 39, 10122 Turin, Italy
e-mail: sara.boninibaraldi@unito.it

C. Salone
e-mail: carlo.salone@unito.it

© Springer Nature Switzerland AG 2020

I. M. Lami (ed.), *Abandoned Buildings in Contemporary Cities: Smart Conditions for Actions*, Smart Innovation, Systems and Technologies 168,
https://doi.org/10.1007/978-3-030-35550-0_8

relations), grouping them by summary and mutual connections in the following three interpretations:

1. Genesis, governance and organisation
2. Business model and sustainability
3. Impacts on the territory.

8.1 Genesis, Governance and Organisation

The first comment concerns the experimental drive and, consequently, the developmental nature of the projects analysed: these are 'emerging', non-planned projects, triggered by the urgent need to respond to concrete—individual or collective—demands by mobilising tangible and intangible resources. They are not the outcome of a set of planned actions but are, in almost all the experiences, a process advancing on a trial-and-error basis. Here the influence of a specific organisational set or business model does not seem decisive, it is rather the need to implement a mechanism which can produce effects moving in the right direction, thus a model which emerges from doing (Mintzberg 1978), according to a process of giving meaning *ex post*.

While in the case of large-scale regeneration works, the conventional sequence of reuse initiatives generally involves the adoption of negotiation logics between private proposers and public administrations and the definition of intervention methods drawn within articulated normative frameworks, cases like the ones we have studied break with the traditional order to move along non-linear paths in an intermediate space between market and non-profit, even though sooner or later they tend to assume an approach that we may define, citing Lindblom (1959), as 'disjointed incrementalism'. In other words, we may state that in these initiatives, the actions are not developed according to criteria of synoptic rationality and rational-comprehensive programming, but rather follow the course of events, learning from their mistakes, changing and improving their course incrementally in relation to the objectives, which in turn gradually change over the course of the works.

Grisù was born in Ferrara, initially driven by a cultural "association of associations" interpreting the collective need to take care of an abandoned public asset, an ex-fire station which was made available free of charge by the province via a loan-for-use contract. The key figures in this first phase are a group of artists pooling their creativity and enthusiasm to recover the asset, but whose different levels of energy find it difficult to come to any agreement over time. After the first experimental phase, the experience quickly changed its skin, turning into a consortium with a strong, individual leadership represented by the current chairman. The explicit aim is to bring life to a "creative factory" through the participation in a regional call for proposals to obtain significant funding to get the operation off the ground. The consortium is led by a Shareholders' Assembly and an Executive Committee elected from among its members, supported recently by a salaried technical director. The members of the Executive Committee receive a small fee for out-of-pocket expenses and the tasks are assigned informally, while the consortium—in line with an explicit work ethic—does not permit any form of voluntary work. The informal organisation of the works of the Executive Committee does not however seem to ensure the required professionalism over time. As the initiative is progressively consolidated, expanding its physical presence in the area, its vocation is progressively shifting towards business activities, however maintaining a broad vision of the consortium's role as a promoter not only of the economic relaunch of a recovered public asset but also of an urban regeneration process. In this *pro-business* evolution, which is also of value to its spatial transformation, a significant role is played by the municipal administration of Ferrara, which is developing ad hoc normative solutions and supporting the project as an outpost for repairing relations between the neighbourhood and the city centre.

A very different route has been taken by Farm in Favara, where the springboard for the reuse mechanism is supported by significant private capital, not particularly interested in any economic return on its investment and driven by strong social commitment. The good relations deriving from the social position of the promoters translate into a clear ability to involve parties outside of the local context. This is a "family" model, with very strong moral values and lying at the centre of a network of extra-local relations which bring contents and cooperation from outside, and which become increasingly important as the experience progresses. The operation has a conventional association-type organisation, in which a dozen salaried employees—not all full-time—and a number of trainees work.

Toolbox, on the other hand, lies within a specifically entrepreneurial framework, born following the bankruptcy of a large company working in the clothing sector. It springs from the desire of one of the ex-owners of the company, to whom the ownership of an industrial building was transferred as part of the share liquidation package. Here a decisive role is played by the size of the initiative: the economies of scale and scope guaranteed by the expansion of the network of project members—freelance workers and businesses renting the premises—produce increasing returns that allow the project to expand further. The organisation is run by a relatively small group of persons, including the manager, who is also the mind behind the initiative, and a few people working in the administrative secretariat (contracts and economic issues), maintenance, *community management* and communication. In legal terms, Toolbox is a limited liability company with single shareholder—the owner, who is also its chairman, while the manager experiments innovative formats and runs the project.

Diametrically opposite to the previous one is the case in Rome, Officine Zero (OZ). This represents the only truly collective project, made possible by a 'negative' opportunity: the closure of a train maintenance company and the subsequent protests and resistance by some activists and ex-employees who occupy the plants, intent on preventing the sale and functional transformation of the premises. Around this nucleus of 'industrial resistance', political thought is developed on the issue of industrial salaried work and freelance work, linked to intangible production, running a number of production lines linked to reuse and recycling. Its governance structure is two-fold: it has an informal line, run by Officine Zero, and another legally constituted line, Zero Off. The operation is simple: everyone bringing their own activities into the premises is part of the Officine Zero project, an informal collective, but not everyone is a member of Zero Off, which on the other hand is an association. OZ works through a general assembly held at varying intervals; participation in the assembly is recommended to all those who orbit around the centre and all kinds of items can be found on the agenda. There is no executive committee with management functions, but there are department meetings (dedicated to various sector activities) and project meetings on more specific issues. Zero Off on the other hand represents the formal extension of OZ, and acts as a legal facility for signing contracts or assigning appointments externally (e.g. workshops for schools) and for drafting contracts, receiving donations etc.

The experience of CAOS in Terni represents yet another variant. This is based on a triple partnership involving a public administration (the Municipality of Terni), a private, profit business (Coop) and other private—profit and non-profit—businesses working in the field of cultural and/or social activities (joint venture). A fundamental role is played by the association Indisciplinarte, which is responsible for guiding the strategies of the initiative. This is the case in which the public sector is most directly involved, both as an investor and in the funding of running costs.

ExFadda in San Vito dei Normanni (BR) is the last experience analysed. This is the only project which is already in its second-generation phase: after the initiator and original 'leader' of the project handed over the reins to some young people, the overall management of the initiative has also radically changed. The governance model includes the Municipality of San Vito, the communication and marketing firm "Sandei" and the cooperative "Qualcuno di diverso". Sandei was granted free use of the ex-wine factory by the Municipality, and subsequently sub-loaned it to the cooperative Qualcuno di diverso to manage all the spaces except the restaurant. The decision-making structure of the cooperative is 'horizontal-meeting' type, and using this form also manages other activities, the most important of which is a 50-ha farm, also loaned free of charge, where olives and grapes are grown.

This is the only experience in which the main legal body—the cooperative—acts both as coordinator of third party initiatives and manager of the project as a whole, and as the direct implementer of social and urban innovation projects and initiatives.

8.2 Business Model and Economic Sustainability

Thought on the business model implies the understanding of how an organisation creates, delivers and captures value (Baden-Fuller and Morgan 2010). In the conventional sense, this means understanding the uniqueness of the value create by single initiatives—their mission, or also their "value proposition"—and how this results in activities for different segments of consumers which generate costs on one hand and revenues on the other. The idea is to strive for sustainability, i.e. the ability to continue to operate over time through the capability of creating recognised value that is perceived as such and is thus able to generate and attract at least as many resources as those consumed.

In this perspective, the first thought triggered by the initiatives analysed is that of the value created. Ultimately, what emerges from all the initiatives is not so much the definition of a single value but rather the stratification of multiple values, a kind of "value mix" which more often than not responds to both strictly productive as well as social, and at times cultural, needs at the same time. What varies is rather the relative weight of the production component over the others. In Grisù, Toolbox and to some extent also OZ and ExFadda, the production component is the main one—in any case combined with a strong drive for social innovation—while the main mission of Farm and CAOS is cultural, with a significant focus on urban regeneration and the social dimension. Thus, the value mix becomes a fundamental component of the business model of the initiatives analysed, clearly reflecting on the definition of the users of the initiatives and the type of activity performed, as well as the economies generated/consumed.

The discriminating dimension in this sense is that of distinguishing between 'internal' and 'external' beneficiaries. In the initiatives based mainly on production activities, the first beneficiaries are in fact those who occupy the spaces and, consequently, actively participate in the project, its governance and organisation. We may think of Grisù, Officine Zero, Toolbox and ExFadda: while the space is open to citizens through public initiatives (some free, some with a charge), or temporary rentals to third parties, the main purpose of the project is to make it possible for and provide support to the persons promoting the initiative and/or occupying the building in question for various reasons. Certainly, there are different forms and types of involvement: in Toolbox, and to some extent also in ExFadda, the governance is centralised, and the residents are mere 'lessees', while for Grisù and OZ the governance is participatory (in the form of a consortium or collective). But in all cases, it is precisely the contributions of the individual members of the collective (Officine Zero), consortium members (Grisù) or lessees (Toolbox) which generate the main source of income (only in ExFadda there is an apparent attempt to support the cooperative's activities also through a specific commercial activity, the management of the bar).

It would however be incorrect to consider the social purpose as insignificant: the impact on the surrounding area, opening up to the neighbourhood and the local people, the social commitments and the provision of services to a broader public, in any case present even in the most specifically entrepreneurial case examined (Toolbox), are elements of fundamental legitimacy, without which the sustainability of the project would be seriously undermined. In other words, the presence of external beneficiaries—who bring negligible or in any case secondary income—legitimize the internal components, which are on the other hand the main source of income. This of course in no way undermines the 'honesty' of the project itself: we do not wish to state that the social commitment is instrumental—or worse, instrumentalised—for other purposes, but it is the fundamental and essential component of a business model based on the joint existence of multiple values. The same applies to some extent also to specifically cultural activities. Let us look at FARM: certainly, without the significant return for the city—also in economic terms, relating to the various commercial activities (B&Bs, restaurants etc.) —it would be difficult to ensure the overall sustainability of the project, in the sense that the legitimacy to act also comes from the economic and image returns of the whole city. In the case of projects developed around a publicly owned asset, the collective interest is what guarantees accessibility to a primary resource, i.e. the availability of spaces at discounted rates. In the same way, for projects developed around a privately-owned asset, the collective interest is what enhances the increasingly cooperative attitude of the public decision-maker. Thus, the concept of sustainability should acquire an extended meaning, with both endogenous and exogenous components, and a significant dimension of recognition and legitimacy. And so perhaps sustainability must be seen beyond mere organisational boundaries, not so much in terms of impact generated but in terms of definition of conditions for the organisations' long-term existence and survival.

Certainly, the significant role of the strict cost/revenue dynamic must be recognised, monitored and suitably guided, otherwise the initiative may not survive. In this sense the attention to cost containment rather than increasing revenue clearly emerges. All the experiences analysed, with the partial exception of CAOS, base their activities on an economy of reuse, recycling, DIY, voluntary work (the latter excluding Grisù), with a creative effort functional to generating value at low cost, using methods and skills that could not be imagined in classic market situations or with direct public intervention. These are all very lean concerns, with very few employees (generally less than 10) or indeed none (OZ). Even in those projects involving high economic investments (Farm, Toolbox, ExFadda) the running costs are generally very low (below 200,000 EUR). This element is highly consistent with the inherent identity of the initiatives analysed, where the sharing of experiences, resources and skills, along with the recovery of the value of abandoned infrastructure resources, becomes a key issue not only for the birth and development of initiatives but also for the experimentation of a new economic and social model—in line with the philosophy of common goods, sharing economy and social innovation—based on values other than neo-liberal or classic welfare state ones. Also on this aspect, CAOS acts as 'control group': however much the efforts of the non-profit partner (Indisciplinate) are driven by the same energy as the other initiatives analysed, the project as a whole is the one which moves the largest investments and current resources, both in absolute terms and in terms of local authority funding. CAOS is the only case in which an agreement with the Municipality involves municipal funding for the management of the project, while in the others analysed there is no annual contribution, only some minor funding for specific calls and projects, which leads to a constant struggle for self-sufficiency (achieved completely only by Toolbox).

Therefore, what allows collective resources and efforts to come together positively to achieve a global vision? What are the key resources and skills, in addition to the material availability of an asset, which make the low-cost generation of value possible? A *fil rouge* among all the initiatives analysed is that of a strong, "charismatic" leadership, a visionary figure able to stimulate, gather and guide the collective contributions towards a common direction. The only truly collective project in the strict sense is Officine Zero: this is certainly the most extreme case analysed, a kind of—benevolent—Hydra—with several heads, in which one of these can be cut off but immediately regenerated, where everyone is important, but nobody is indispensable. Here too, however (and only apparently in contradiction with the type of initiative in question) the entrepreneurial element is crucial. In all cases, these are highly entrepreneurial initiatives, understood as activities in which "someone" takes a high risk and invests his own capital and resources, putting significant creative and innovative forces into play: a kind of new civil entrepreneurship placed beyond market logics and characterising a new form of post-political citizenship (Salone et al. 2017). In this logic, relations are important on several levels: in activating and stimulating the initiative of individuals, volunteers, members; in managing the delicate relationships with the public administrations; in attracting attention, resources and skills from outside; and in the dialogue among peers. Social entrepreneurship becomes a kind of community of practice, with similar logics and values: a close community characterised by similar values but also specific, real and virtual meeting places, not only and not so much at local level but rather at national and partly international level. Generally, the focus is therefore on an economy based on relationships, in which the true key is that of activating others via the self, and the self via others.

The real challenge for sustainability is thus not (only) that of generating sufficient revenues to cover costs, but to continue to activate their own and others' efforts in a common direction. This direction can, indeed must, change, in fact precisely because sooner or later the key internal factors (the composition of the collective, the leader) as well as the external factors (the municipality, the contract terms, and so on) change or must necessarily change. The various development paths thus bear witness to the search for a delicate balance between adhering to the original project and the capacity for transformation, with the need for a resilient approach in which the only constant element is the lack of certainty. A clear attempt to innovate on traditional models emerges from the initiatives analysed. In the case of OZ in Rome, the acquisition of the new space in alternative to the original location implies the re-thinking of the very mission of the project and a radical redefinition of the organisation chart; in ExFadda, it is the introduction of the farming activities; or again, in FARM, it is the re-invention of the limited company, also structurally becoming a community business (Mori and Sforzi 2019). Of course, long-term sustainability is still to be demonstrated. What clearly emerges however is that the apparent simplicity of the situations analysed on paper reveals a multitude of heterogeneous elements which make their interpretation far more complex. This is very clear when looking at the legal form. Many of these projects have a two-fold nature: both limited liability company and association (CAOS, FARM), both collective organisation and association (OZ), both limited liability company and cooperative (ExFadda). This two-fold nature seems functional to the search for innovative formulas and creative economic and normative solutions. Currently, the legal framework in any case has no ready-made solutions, and this aspect must be appropriately developed in future with the contribution of the legal culture.

8.3 The Socio-spatial Relations Generated: From the Relationship with the Urban Space and Its Actors to Supra-Local Relations

At the current state of knowledge emerging from the field investigations, it is rather difficult to define one or more general schemes that can describe the spatial behaviour of the organisations and the effects of their activities on the space, understood both with reference to the local context and in relation to the supra-local networks activated. We can at most indicate some specific features emerging from the interviews and field observations, trying to suggest possible pathways and scenarios for development starting from the actual situation. The first two features concern the relations with the 'local' and the relations with supra-local networks, while the third concerns the relations with local public powers and their logics.

– *The redemption of the 'local' as a trigger for action*

The relationship with the urban context can be interpreted in two ways. The first and most immediate, but also the most 'banal', relates to the use these organisations make of urban materials: abandoned industrial buildings, empty, degraded residential structures, public buildings stripped of their function. All these physical elements of the urban scene are returned to a value-making circuit which initially appears as a process of cultural and social reinvestment and, potentially, in a more mature phase, an economic value-making circuit.

The second way concerns the socio-spatial relations which more or less effectively underpin the local context but which may also be extended to 'higher' scales. In fact, as explained below, the scales of action mix together in a fluid dimension in which almost all the actors studied move from the local to the trans-local, adopting spatial behaviours which are fairly common today and which change the very role of the architectures around which the projects are implemented, according to the principle of *spatial agency* as defined by Awan et al. (2013). Buildings are not only material objects subject to rules of use and intervention methods but are also part of socially rooted networks, producing consequences which go beyond the architectures themselves, making them 'protagonists' of transformation processes along with the social actors who return them to the circuit of use with 'political' (in the broad sense) intentions.

In fact, the beginnings of almost all the experiences are linked essentially to the emergence of projects aiming to respond to unsatisfied needs within the range of local cultural, social and professional services. From the cultural associations behind the establishment of the Grisù Consortium in Ferrara to the commitment to cultural animation of FARM in Favara, from the physical and functional requalification of the abandoned industrial buildings in Terni where CAOS is located to the antagonistic militancy of the Officine Zero collective in Rome, and the social entrepreneurship of the group leading the recovery of ExFadda, all these experiences start from local stimuli and continue on a trajectory in which their cooperation networks expand outwards. The only exception seems to be Toolbox which, perhaps due to its specifically entrepreneurial nature, does not seem to be particularly bound by close relations with its socio-spatial context.

Local relations also stand out very clearly in the declarations of principle, even though subsequently they are marked by difficulties arising with part of the community which does not recognise the 'rooted' nature of the experiences (in Terni as well as Ferrara), or towards public interlocutors which consider them to be politically connoted (again Terni and Ferrara).

– *Towards an extension of the scale of spatial relations*

The consolidation and development of the initiatives bring an expansion of the cooperation networks to at least regional scale, for reasons related to the expansion of the business activity—search for new customers, association with other parties in order to be competitive and participate in public competitions with the necessary strength (Grisù in Ferrara), increase cooperation to enhance the cultural services and de-provincialise the local atmosphere (FARM in Favara, where national and international relations concern the training activities involving the Polytechnic of Milan and "resident" foreign artists). Also in the case of ExFadda, the commitment to revitalising the local scene according to the principles of social entrepreneurship and civil economics, which the promoters of the initiative firmly believe in, has led to the creation of increasingly complex relational networks in time and in space, involving subjects from outside the local context and creating contacts with a network of organizations working according to similar methods and purposes even further afield. This development can also be seen in the more informal experience of Officine Zero in Rome, where militants occupy the company headquarters aiming to "remain" and "resist", contributing to the redemption of Casal Bertone and—even—Roma Est, thus also having to negotiate with new global interests, breaking the classic bonds of localist antagonism.

– *Relations with local powers*

These experiences have several types of relations with the public sector: from cooperative or conflictual dialogue with local powers to the recourse to public funding, inclusion in knowledge-exchange networks and public-guided expertise, the forms of relations may change substantially and develop according to very different patterns.

Generally, we can state that no dominant behaviour emerges, and the rather fluid interaction depends greatly on the attitude of the local administrations. In fact, if we observe the relations established with the local powers in the strict sense (substantially the municipal administrations), in some cases cooperation has always been the dominant feature of the experience (Grisù in Ferrara and ExFadda in San Vito), while in other situations the municipalities were initially indifferent or hostile but were then in some way forced to follow the process with more attention (OZ in Rome and Farm in Favara). Again in other situations, the administrations were among the promoters, but then ignored the initiative for reasons essentially linked to political circumstances in the local government.

The experience of Grisù in Ferrara also shows how even generally cooperative behaviour can however periodically be undermined by bureaucratic rigidity, often leading to seek ad hoc solutions within the boundaries of legality but with broad margins for 'administrative creativity' to overcome normative difficulties, which is possible only in the presence of 'enlightened' administrators and civil servants.

In Favara, on the other hand, initially the municipal administration was wholly indifferent, and then indeed became hostile, accusing Farm of illegal occupation of public land during the set-up of an artistic installation in a courtyard in the historical centre of the town. Following meetings held to solve the question, the Municipality now understands the positive significance of the project for the development of commercial fabric and hotel services in the historical centre and declares it of public interest, but in concrete terms it does not support Farm's actions.

Officine Zero in Rome is certainly the most different and, in some aspects, extreme experience: initially the municipal administration which holds jurisdiction over Casal Bertone did not want to become involved as the occupied asset is privately owned. The refusal of the Administration to take a political stand was stigmatised by the occupants, who then lobbied the Region, where they could count on more sensitive administrators who played an intermediary role in the negotiations between Officine Zero and the owners of the area, Paribas—Banca Nazionale del Lavoro. Although not responsible for urban disputes, the Region accepted this role and eventually succeeded in involving the Municipality, which in the meantime in 2018 had sent an eviction notice to the occupants. Here the most interesting aspect is the recognition of the role of interlocutors not by the administration, which in fact kept a 'safe distance' from the claims put forward by the social parties, but rather by the owners, an international global finance institute which demonstrated far more familiarity with the management of conflicts than the—practically no—familiarity of the local public powers.

The links with the public on the other hand appear substantial and fundamental for CAOS in Terni, where a political leader, an 'enlightened' Municipal councillor, found 'creative' solutions in the current normative and financial framework. The call for proposals was worded pertinently, adapting perfectly to the needs and cultural programme of the applicants. The second call the association was able to participate in was on the other hand the result of the next administration, with a much less happy outcome in terms of both contents and allocated resources. The situation was further complicated by the recent change in municipal administration, which does not guarantee the survival of the project.

Finally, the events of ExFadda demonstrate yet another cooperation model. In this case the cooperation with local public powers guarantees not only a considerable continuity and progressive expansion of activities, but even contemplates the co-design of public policies: indeed, when works began to define the urban regeneration guidelines for some parts of the town of San Vito dei Normanni, the Municipality wished to include the representatives of ExFadda in the working group, composed also of experts from outside the local reality.

8.4 Final Consideration

To conclude this brief analysis, it is worth reflecting on the ownership of the assets in question, checking if and how this differentiates the cases in terms of governance, legal form and organisation on one hand, and mission, business model and sustainability on the other. Three of the six cases analysed concern publicly owned buildings (Grisù in Ferrara, CAOS in Terni and ExFadda in San Vito dei Normanni), while three are private properties (Farm in Favara, Toolbox in Turin and Officine Zero in Rome). This choice was made by the researchers at the start of the project, to guarantee a certain diversification among the cases; some useful reflections can perhaps in any case be made.

Despite the diversity of the cases, the link in the first aspect—the relationship between the ownership of the asset and the governance choices—appears quite clear. In the case of Grisù in Ferrara for example, the choice of consortium (to the detriment of the previous association) was dictated by the need to respond to a public competition issued by the Region for the implementation of the creative factory. CAOS is the most emblematic example of a public initiative, with the creation of a specifically established public-private partnership. ExFadda on the other hand is the most complex case, where the link between the ownership of the asset and the legal form of the management organisation, as well as the form of governance, is less clear: from an initial involvement of a for-profit business (Sundei s.r.l), to a management subcontract to a type B social cooperative ("Qualcuno di diverso"). In any case, all three realities were created from a public competition, and therefore have to deal with the formal constraints imposed by the call. More interesting is the genesis: despite the public ownership of the asset, both Grisù and ExFadda come from mainly grass-roots operations and the initiative of private organisations (a non-formally established group of artists and citizens in the first case, the manager of a communications company and a project manager in the second). Despite this, the public sphere is certainly involved in the development process and the definition of project contents. The current identity of Grisù (as incubator and accelerator of cultural and creative production activities), ExFadda (social innovation and urban regeneration), and indeed CAOS (cultural production centre), is the result of continuous dialogue between public and private, seeking to find a synthesis between at least partially converging interests.

The public ownership of the asset is not however a guarantee of economic commitment by the local administration. Neither ExFadda nor Grisù—beyond the (minimal) initial investment for the restructuring of the properties—receive any stable funding from the public administration. The only case in which the public contribution has been of fundamental importance for both infrastructural investments and day-to-day management is CAOS. This is an important element, if we consider the fragility of the relations with the local administrations: apparently the "safest" reality, CAOS is in fact the most fragile, the one that can most easily be compromised by external elements (a hostile administration, a local economic crisis, a change in the priorities of local policies).

Observing the three privately owned assets, what initially emerges is the extreme heterogeneity of the initiatives analysed: a for-profit business in the case of Toolbox, a family-run social project in the case of Farm, a collective and political project in the case of Officine Zero. In the first two cases—and despite the different legal solutions identified—the ownership and the management organisation in fact are one and the same. In both cases the project is univocal, implemented in an asset owned by the very party setting up the initiative. The fact that the asset is owned by an investment fund, and the project managed by an association or company with a separate legal personality, is indeed secondary. This leads to the total and direct involvement of the asset owners in both the definition and development of the project and in the investments required to implement and manage it. In both cases the economic commitment is clear in both the initial purchase and restructuring of the premises and in the support to the day-to-day management, at least until the activities are up and running (achieved only by Toolbox).

In the case of Officine Zero on the other hand, the ownership and management organisation are on opposite sides of the barricades: this is indeed an act of forced occupation of a privately-owned asset. It is clear that in this case no economic support is guaranteed by the ownership. It is not however true that the owners of the asset have no influence in terms of retaliation on the form of management or business model: where none of the active subjects is the owner, the force counteracting that of the ownership lies precisely in the definition of a governance and sustainability model based on a medium-large scale horizontal participation by the occupants.

Ultimately: the owners of the asset always appear to play a significant role in the governance choices and the project contents of the initiatives analysed. They do however seem to be a guarantee of economic support only where the ownership and management organisation are one and the same (Toolbox, Farm and CAOS). In all other cases, the link between the ownership and economic responsibility is considerably weakened (Grisù and ExFadda) and even non-existent (Officine Zero). This clear element emerging from our case studies cannot but lead to interesting questions on the delicate (in)coherence of the business models developed and their long-term sustainability. On one hand—and this make sense above all for publicly owned assets—we may state that either ample freedom is given to define projects and actions, at this point also delegating the economic responsibility for the project, or the project is bound in legal and content terms, also, at least partly, taking responsibility for all related economic costs and benefits. And perhaps the freedom of these projects to experiment "outside the box" from various points of view is due precisely to the lack of public funds, which bring with them a whole series of constraints and rigidities which in the cases analysed are still fairly limited. On the other hand, however, how far the constraints posed are sufficiently "flexible" and how far the parties promoting these initiatives have the effective strength, skill and energy to stand on their own feet, are still open questions. Certainly the link between decision-making freedom and economic responsibility is an important element—and partly a winning feature—of the private initiatives, and is an issue on which policy makers should also appropriately reflect. Some words, in the end, deserve to be said concerning the relationship with the socio-spatial context. Even if all the experiences entail a quite obvious function of redefining or redesigning the context where they are located and act, and are aware of the influence they can have directly or indirectly on it, only a few of them recognise a clear commitment in a regeneration process, namely Farm, ExFadda and Officine Zero.

However, also in these three cases the actual engagement is practiced in different ways, accordingly to their peculiar origin and history. Farm considers the betterment of the architectural quality and use of the historical centre of Favara as a clear purpose of its mission, and holds relations with external partners including universities, architects and artists. ExFadda takes part in the local planning initiatives for the urban regeneration of San Vito dei Normanni, cooperating with the Municipality. Officine Zero claims that its experience should be replicable in other derelict parts of peripheral Rome, striving for ecologically sustainable regeneration and re-appropriation of common land (even though the land it operates on is private property), but relations with the Municipality remain conflictual.

References

Awan, N., Schneider, T., & Till, J. (2013). *Spatial agency: Other ways of doing architecture*. New York: Routledge.

Baden-Fuller, C., & Morgan, M. S. (2010). Business models as models. *Long Range Planning, 43*(2–3), 156–171.

Lindblom, C. E. (1959). The science of muddling through. *Public Administration Review, 19*(2), 79–88.

Mintzberg, H. (1978). Patterns in strategy formation. *Management Science, 24*(9), 934–948.

Mori, P. A., & Sforzi, J. (a cura di). (2019). Imprese di comunità. Innovazione istituzionale, partecipazione e sviluppo locale, Bologna, Il Mulino.

Salone, C., Bonini Baraldi, S., & Pazzola, G. (2017). Cultural production in peripheral urban spaces: Lessons from Barriera, Turin, Italy. *European Planning Studies, 25*(12), 2117–2137.

Part III
Discussion and Conclusions

Chapter 9
Shapes, Rules and Value

Isabella M. Lami

Abstract The issue concerning unused buildings revolves around the fact that they constitute a huge real estate asset for which there is no longer a demand. However, this is an asset that has been subject to significant investment and, therefore, three crucial questions spontaneously arise: (i) Can these abandoned assets still be useful? (ii) What does it mean to enhance their value? and (iii) What are the suitable conditions for launching innovative development processes? This chapter provides a possible answer to the three key questions.

9.1 The Issue

The issue concerning unused buildings revolves around the fact that they constitute a huge real estate asset for which there is no longer a demand. They are, apparently, useless. In Italy, the GDP has stalled and population growth is inexistent (European Commission 2019; ISTAT 2019).

However, this is an asset that has been subject to significant investment and, therefore, three crucial questions spontaneously arise: (i) Can these abandoned assets still be useful? (ii) What does it mean to enhance their value? and (iii) What are the suitable conditions for launching innovative development processes?

The answer to the first question is positive, namely, they may be useful if the appropriate conditions are implemented. Buildings still have the potential to produce value, particularly in big cities. This statement is justifiable from various points of view, with a particular relation to a cultural environment that enhances the concepts of circular economy (Kirchherr et al. 2017) and adaptive reuse (Plevoets and Van Cleempoel 2013; Robiglio 2016), by choice or by necessity, whichever the case may be.

I. M. Lami (✉)
Department of Regional and Urban Studies and Planning (DIST), Politecnico di Torino, Viale Mattioli 39, 10122 Turin, Italy
e-mail: isabella.lami@polito.it

© Springer Nature Switzerland AG 2020

I. M. Lami (ed.), *Abandoned Buildings in Contemporary Cities: Smart Conditions for Actions*, Smart Innovation, Systems and Technologies 168,
https://doi.org/10.1007/978-3-030-35550-0_9

The answer to the second question is more intricate, and it may be connected to the concept of the sharing economy (Botsman and Rogers 2010). An initial, immediate response may be that when an abandoned building starts to be reused, its exploitation has already been successful. A slightly more detailed answer would require a distinction between private and public abandoned buildings. For private buildings, the discussion could be limited to the profitability of the urban renewal; for public buildings, various scenarios may apply, due to the consequences of a public decision in a context of allocating scarce resources and public responsibility towards the civil community. A third type of reaction could be to discuss the types of value that can be created (i.e. not just financial value, but also social and cultural value, meaning values per se and not as leverage to increase the financial value of the recovered asset).

As to the last question, it seems that the challenge faced by today's cities is to redefine their priorities. Rather than managing themselves according to codified rules or focusing only on the efficiency paradigm concentrated on revenue/expenditure or cost and profit management, they must reflect on the optimal combination of both, hard and soft infrastructures (Landry 2008; Sacco and Tavano Blessi 2009).

In this chapter, these aspects will be critically discussed, paying particular attention to the Italian situation and drawing some insights from the six analysed case studies.

9.2 Is Unused Asset Synonymous with Useless?

To answer to this question, two key concepts must briefly be introduced: Circular Economy and Adaptive Reuse. They both explain the perception of the abandoned building as a potential asset.

9.2.1 Circular Economy

The traditional linear economic model "take-make-dispose" based on the possibility of having access to large amounts of resources and energy is a model that can no longer be pursued. Modern cities produce 50% of global waste and 60–80% of gas emissions and they consume 75% of global natural resources (Ellen MacArthur Foundation 2017). McKinsey (2016) reports that between 2015 and 2030 there will be a 91% growth in city consumption. Focusing more specifically on the problems at European level in the building context, three issues arise: firstly, buildings continue to use enormous amounts of energy; secondly, most buildings remain empty and unused; and, thirdly, the demolition and construction of buildings accounts for 25–30% of the waste generated and their recovery is not often an attractive option due to the fact that the materials are contaminated (Ellen MacArthur Foundation 2013).

In this context, an economic model based on sharing, reusing, repairing, restructuring and recycling in a closed circuit which seeks to maximise the value and utility of products and materials has been developed: the "Circular Economy". Williams (2019, p. 5) claims that the Circular Economy "*is not merely seen as a preventative approach, but as an ecologically restorative and regenerative approach, repairing previous damage by designing better systems within industry*".

The Circular Economy model is based on three principles (Ellen MacArthur Foundation 2015): (i) preserving the natural capital of finite resources by selecting technologies and processes which use renewable resources, and by creating conditions for soil regeneration; (ii) maximising resource yields by planning a loop circulation of products through regeneration, renewal and recycling so that materials can continue to contribute to the economy; (iii) designing the management of negative externalities, such as the air pollution, water and noise, release of toxic substances and climate change, in order to reduce damage and promote the system.

In order to adopt a circular economy within the built environment, it is worth mentioning some relevant aspects of the urban context (Williams 2019). First, land is the most important resource we have in the city, so in a circular economy any wastage of this resource should be limited; its recycling should also be facilitated so as to optimise the use of resources. Second, urban infrastructure is a mine of resources, representing 50% of global raw resources consumed annually and 33% of total energy consumption and emissions; the reuse of such resources is fundamental and should be encouraged in a circular city. Finally, some physical structures in our cities are obsolete and in need of rehabilitation or replacement. Demolition and renewal have implications in terms of resources; therefore, to limit waste, the Circular Economy encourages reuse and the use of modular and flexible infrastructures to facilitate their adaptation.

Accordingly, the next concept, Adaptive Reuse, is closely related to the concept of the Circular Economy as a tool that encourages the circular model.

9.2.2 Adaptive Reuse

In a study on the conditions of actions for enhancing the value of abandoned buildings and unused spaces in cities, the concept of "Adaptive Reuse" is crucial.

After several theoretical discussions on the conservation of historical buildings, in the 20th century Adaptive Reuse came to be defined as a creative discipline with its own rights and theories in favour of the preservation of cultural heritage (Plevoets and Van Cleempoel 2013). This concept enabled cities to cope with the increasing amount of abandoned assets over the years and, in parallel, to address huge social, technological and environmental changes (Douglas 2006). Therefore, Adaptive Reuse emerged firstly as a practice and then as a theory (Robiglio 2016) based upon the assumption that buildings, areas, districts and sites are not static entities designed simply for one single use during their life cycle. This approach can be defined as the practice of introducing a new content in an existing container (i.e. building, infrastructure, area), paying particular attention to the needs of society and following the principle of maximum conservation and minimum transformation (Robiglio 2016). As a theory, this phenomenon has been described using different terms: conservation, refurbishment, rehabilitation, renovation, remodelling, restoration (Douglas 2006). Even though those terms are often used as synonyms, it is worth noting that they relate to a broad spectrum of possibilities which differ starkly from each other. More specifically, an ascendant scale should be envisaged which starts with the basic work of conservation and ends with the total reconstruction of the building and, between them, the refurbishment, rehabilitation, renovation, remodelling and restoration actions are distributed with an increasing degree of intervention. In general terms, Adaptive Reuse is an approach that should not involve heavy work and changes to the existing building, but it is a strategy to ameliorate disused buildings, transforming them into new entities with new purposes (Bullen and Love 2011a). While, in the past, the concept of Adaptive Reuse usually involved industrial sites (Langston 2010; Günçea and Mısırlısoya 2015), recently many interesting examples can be found in historic districts (Bullen and Love 2011b; Plevoets and Van Cleempoel 2011), highlighting the need to recover heritage buildings, viewing them as opportunities rather than risks.

All buildings that have lost their function become obsolete, but they represent a value or a symbol for the place or the community: they acquire social, cultural and emotional value over the years, developing an intrinsic memory, which is difficult to erase without opposition (Fiorani et al. 2017; Conejos et al. 2014).

From an economic perspective, extensive literature confirms that the adaptation of a building or an area is cheaper than implementing a new project (Douglas 2006; Bullen and Love 2011b; Remøy and Van der Voordt 2007; Kohler and Yang 2007); however, an economic reflection on the reuse of existing buildings must be included, above all, in a Circular Economy perspective, where reuse is encouraged, as opposed to demolition and reconstruction, as it reduces the consumption of raw materials and energy used in the process, lessens waste and preserves portions of the urban landscape.

In this context, Adaptive Reuse can be considered a valuable approach for the new sustainable rebirth of the city (Dewiyana et al. 2016) and, in order to achieve a positive result in the transformation process, an in-depth analysis of the specific characteristics of the sites and buildings—both positive and negative—must be carried out. Accordingly, Robiglio (2017) identifies the crucial points that a helpful analysis should follow. First of all, the location plays an important role: we should investigate the accessibility, connections, services of the areas and visibility, in order to be aware of the opportunities of the project sites and the potential future inflow. Secondly, the scale of intervention, which is defined on the basis of the context and goals of the project, is important. For the reuse of a whole urban section, often the process starts from revitalising one of its areas or buildings, thus becoming the first step of the entire reuse project. Thirdly, it is fundamental to ascertain the potential of the site or building, which must be maximised, both for the new activities and to preserve the link with the past. Finally, in order for the project to be successful, the local and global interests must be investigated, in order to present a bold vision, which also responds to market demands and society's lifestyle changes. Having identified a feasible transformation method, partners should be involved in the project from the city scale to institutional level, and the renown of the abandoned and forgotten place must be increased among the communities. In this regard, temporary uses or art could be a powerful springboard for this effect: they are generally inexpensive and they create business opportunities, and, above all, they can activate unused areas and change the negative perception of sites as being blighted and neglected towards a positive vision, increasing their visibility and activity within a neighbourhood (Németh and Langhorst 2013).

It is worth mentioning that, as the activation of this reuse process involves different political, economic, social, environmental and technological factors, this will inevitably involve a complex decision-making process due to the participation of numerous stakeholders having different objectives, concerns and interests (Abastante et al. 2020; Abastante and Lami 2018).

9.2.3 Unused Is Not Synonymous of Useless

Theoretically, every asset can and, according to the above concepts, should be renewed. Through smart actions, all buildings can be reused, with the capacity to rebrand and reframe the containers, but it is obviously necessary to ascertain if the conditions exist to make the intervention worthwhile. These conditions include "traditional" ones, considered fundamental in the real estate assessment (Orefice 2007):

1. Location characteristics, which can be summarised as follows: location of the property with respect to the urban centre; quality of road connections and public transport services; presence of collective equipment (schools, etc.); presence of commercial activities; quality of the external environment (site amenities, absence of pollution, presence of green spaces, etc.);
2. Typological features, including: age of the building; architectural features (to be assessed both in relation to the qualities conferred and in terms of any higher maintenance costs as well as restrictions on use or modifications); type and static conditions of the structures, roofs and finishes; compositional characteristics; characteristics and conditions of technological systems;
3. Production characteristics, which are linked to the legal and economic conditions in which the real estate is located or which may arise within a reasonable period of time in the future. These are: current rental situation, presence of easements, rights, charges, encumbrances; legal status of the property, etc.

While location is fundamental in the same way as in any real estate operation, in an intervention of adaptive reuse for abandoned buildings, the second point plays a crucial role (namely typological characteristics, which can be synthesized with a single word "the shape"). Some places are more complex than others. For example, industrial buildings, a classic case of buildings in need of reuse, often have characteristics of greater adaptability, given their large size and broad spans (see the chapter by Baima and Robiglio).

As for location, the Toolbox case study (Turin) is emblematic, as it is well-connected in terms of public transport and is not far from the two Academies of the city, the Politecnico and the University of Turin.

As for typology, it could be assumed that the aspect which made Factory Grisù (Ferrara) a critical regeneration is the building chosen by the developers: there are too many walls, generating small rooms. However, the Ex Fadda regeneration in San Vito dei Normanni works perfectly as the building presents a large space that lends itself well to different uses: it could be used as a single large open space for events or as a theatre, or it could be divided into small rooms for different activities.

As far as the *production characteristics* are concerned, profound differences with respect to the transformation process and the outcome of the operation are linked to the ownership of the building and the possibility of using it on loan through controlled rent (see the chapter by Mangialardo and Micelli).

9.3 What Does It Mean to Enhance the Value of the Unused Asset?

Three types of interpretation can be identified and thus three different answers can be given to this question.

An initial form of value enhancement could be seen as the transition from a status of non-use to one of use. This interpretation is intrinsically linked to the concept of the Sharing Economy, meaning "an economic system based on sharing underused assets or services, for free or for a fee, directly from individuals" (Botsman and Rogers 2010). It entails the activity of sharing, exchanging or renting resources without the need to own the goods. The Sharing Economy is a growing model based upon access to resources and peer-to-peer relations facilitated by online digital platforms. It is worth mentioning that the sharing economy should lead to a change in the physical character of the built environment for various reasons. The Sharing Economy is no longer tied to a fundamental concept of traditional cities or physical proximity, but the new peer-to-peer services based on virtual proximity will replace current commercial activities. Furthermore, the transport sector will strengthen connections even in areas that are currently under-served, changing relationships between the home and the neighbourhood, and leading to the emergence of new neighbourhoods. Finally, the residential micro-units for housing use located in dense urban centres with services will no longer be needed and desired by potential users, since they will have access to assets with larger spaces through the Sharing Economy without having to own them (Davidson and Infranca 2016).

However, while, in light of this concept, use is still better than non-use, this interpretation is so broad that it can also include illegal forms of reuse, as in the case of Officine Zero described in the book. It must be specified that there is no *ipso facto* appreciative dimension to these phenomena, but only the intention to illustrate the wide range of possibilities of reuse.

A second type of interpretation of the enhancement of value concerns the distinction between private and public. According to this interpretation, three aspects are fundamental and, more specifically, the private or public nature of: the owner of the abandoned building, the developer who initiated the reuse process, and the funder (see the chapter by Moroni et al.). While the reasons why public and private assets remain unused are more or less the same, linked to access to capital, this distinction has considerable implications, including the objective of the operation and the indicators to be used to verify its success. Whereas, for a private developer, enhancement can range from just increasing the market value of the asset to the willingness to promote forms of social innovation, in the case of public urban regeneration, the social responsibility is obviously much stronger, and these places of reuse create value especially when they become platforms that enable social change. In the cases examined in the volume, Toolbox is a 100% private project having a business logic through experimental forms of economy; Farm Cultural Park sees a private investment that promotes social innovation mainly through culture, with effects that are not limited to the properties directly involved in the reuse, but concern firstly the neighbourhood and then the entire city. On the other hand, there is Chaos, which involves a massive public investment and a perhaps excessive intervention size which is struggling to find a long-term sustainability dimension.

A third and final answer to the initial question may involve a discussion on the types of value that can be created: not only financial value, but also social and cultural, intended as values per se and not as leverage to increase the financial value of the asset recovered (see the chapter of Mecca and Lami). According to Sacco (2017), it is entirely inappropriate to think that the economic enhancement of culture is a safe bet. The exact opposite is true: to create development via culture, it is necessary to enter fully into a logic of investment, risk, experimentation, knowing that culture makes sense—even economic—only when it is alive, capable of exploring new territories, inducing change, and releasing new energy.

In this context, Throsby (1995) defines the term culture with two interpretations: firstly, culture can be seen as a set of activities; secondly, culture is a set of practices and attitudes fundamental to the functioning of societies. Thus, culture "is expressed in a particular society's values and customs, which evolve over time as they are transmitted from one generation to another" (see p. 202 in Throsby 1995). A historic building embodies history, tradition and ties a community together, so culture brings with it the notion of "cultural value" (Throsby 1999). Therefore, a cultural asset contributes to cultural value.

Throsby (1999) describes cultural capital as the cultural value embodied in a good and these goods can be both material and immaterial. Based on this, buildings, structures and places seen as tangible and material goods carry a cultural meaning, more commonly known as cultural heritage. These assets contribute to producing the community's future cultural capital. Intangible assets are, on the other hand, represented by ideas, traditions and values that identify a group of people, such as literature and music.

The driver in reuse operations of abandoned buildings is often culture, which is a driving force and a fundamental capital (see the chapter of Bonini Baraldi and Salone). It was the trigger in Grisù, when a cultural association "resurrected" the former fire station, channelling the interest of the community living in the surrounding areas and having the support of the Public Administration; it is the essence of Farm Cultural Park, to the point of making this small Sicilian town visible in the international artistic debate; it is the strong connotation of Caos; it is the common denominator in Ex Fadda due to the varied activities carried out by citizens within that one container.

Finally, an aspect well-described by Sacco (2017) must be emphasised. The future of cultural assets as a possible development platform travels along very different paths from those usually evoked in debates on the subject: production and creative entrepreneurship, the ability to connect to major international networks, the ability to attract investment. In other words, they are guidelines for development that require genuine entrepreneurial capacity, an effort that requires ideas, skills and, above all, credibility. In order to carry out this revolution—which is now as necessary as it is urgent—territories must become candidates to be laboratories for the development of good practices to be progressively extended to the entire nation.

9.4 What Are the Conditions for Actions?

In zero-profit areas, facing a radical restructuring of the hierarchical relationships between places, a new attitude is needed with respect to interventions to reuse the building stock. Attention should shift from the final outcome of the restructuring operations (quantity, use, morphological aspects, etc.)—an attitude typical of traditional urban planning—to conditions that can activate these operations: not conditions that stifle them, but policies that support them.

This represents a normative and cultural change, where the rules are less focused on the material processes of producing space and more aimed at fostering the construction of relationships, and the space will then follow. It was once the places that determined movements, now it is movements that determine the location. It is a cultural challenge not only for Public Administrations, but also for entrepreneurs, who are often not well-established in the real estate sector but must develop new skills.

If the conditions for actions involve developing enabling policies, we must attempt to define the main factors of those policies.

The starting point could be Landry's reflection: "cities now compete by harnessing their tangible (as well as hard and intangible) soft asset base and project and orchestrate this iconically. Very few cities understand this" (see p. xviii in Landry 2008). It requires the comprehension of new competitive urban tools, such as: the networking capacity of the city; its cultural depth and richness; the quality of its governance, design awareness and how to use symbolic and perceptual knowledge and eco-awareness.

In this sense, Sacco and Tavano Blessi (2009) emphasize the importance of finding a careful balance between the physical and intangible components of cultural initiatives, i.e. between cultural 'software' and 'hardware' affirming "whereas the latter provide relatively immediate benefits in terms of economic impact and of endowment of usable physical capital, it is equally necessary to pursue carefully selected programmes of initiatives that involve and challenge the local community at various levels, fostering a social attitude based on intrinsic motivation rather than on instrumental opportunism" (see p. 1115 in Sacco and Tavano Blessi 2009).

As highlighted by Ferilli et al. (2011), a culturally active environment, when adequately supported by a sufficiently large level of social capital (Swart and Kinnie 2003), fosters the emergence of organisational environments that are open to and ready for change. This paves the way for the individual and collective accumulation of cultural capital (Throsby 1999), thus stimulating a positive growth dynamic, such as more skilled workers interested in further investment in intangible capital, but also the attraction of more financial investments and venture capital.

In this perspective, human resources are complementary to natural resources of competitiveness.

Landry (2008), when discussing the creative city, states that there are four groups of elements to assess:

1. *Material things*: natural setting and location; natural resources; history; heritage and the built fabric; infrastructure (transport, housing, ICT, etc.);
2. *Activities*: urban maintenance, from waste collection to social care; traditions (both tangible and intangible); the presence of industries and services; levels of skills and talent; and the range of activities, such as artistic and community-based festivals and events;
3. *Matters of attitude*: level of openness, tolerance, joined-up thinking, entrepreneurship; the perceptions of the city internally and externally; and the presence of a culture of curiosity, creativity and competence;
4. *Perception and organisational concerns*: organisation, management and delivery, such as empowering individuals and companies; strong partnerships, as well as rethinking the regulations and regimes of incentives.

Based upon the analysis of the six case studies illustrated in the book, we can highlight the complementary presence of some hard and soft infrastructures (which could be included in the points of the above list), which seem significant in giving rise to successful reuse processes of abandoned assets.

In the six case studies analysed in the book there is always a crossover between necessities and vision (which could be included in points 1 and 3 of the above list): people who need space but lack financial availability, and people who are dreaming of a new society, a new form of working, culture and activities. In investigating the soft and hard aspects, vision could be considered as a soft skill which must not be underestimated. This vision can be expressed as the dream of a new way of working presented by Officine Zero in Rome or the dream of Farm Cultural Park of a new society that involves people in cultural activities, or, moreover, the dream of Ex Fadda of a collaborative community, which develops new activities.

The second crucial complementary relationship is the good location (in relation to the urban fabric) and the presence of an Institution that produces cultural capital on a regular basis (typically a University); which could correspond to points 1 and 2 on the above list. In this sense, the weak point of the Factory Grisù case study in Ferrara is probably the fact that it is not central and is not connected to University.

The third relationship is between architecture and perceptive and organisational concerns (points 1 and 4), through a transversal view that is able to capture not only the buildings, the full and empty spaces, but, above all, the system of multi-purpose relationships that is established between practices, populations, and built space. Buildings/spaces that seem to be successful, i.e. that are capable of being lively and, consequently, also safe, are spaces that lend themselves to multiple uses. In this sense, architecture has the connotation of hardware, designed and built to facilitate connections and actions, but it also becomes a soft element, as a device for the coexistence of collective experiences. In this regard, Farm Cultural Park is emblematic. Another example of best practice is Ex Fadda, an initiative responding to the lack of locations for the creation and expression of youth and, at the same time, to enhance the contribution of new generations to this resource for city regeneration.

Directly related to the last topic is the issue of funding, overlapping with all four points of the above list. In this sense, an excellent model is that of "Bollenti Spiriti" of Apulia Region, thanks to which Ex Fadda was established (see the chapters of Porta and Abastante and Bonini Baraldi and Salone) within the "Laboratori Urbani" (Urban Laboratories) programme.

Laboratori Urbani is the action of Bollenti Spiriti to recover abandoned public buildings to be transformed into spaces for young Apulia inhabitants. Through a public call addressed to the municipal administrations, in the period 2007–2010 the Apulia Region funded 71 projects submitted by 169 municipalities, to recover 151 underused buildings owned by them and to transform them into "Urban Laboratories" for young people (bollentispiriti.regionepuglia.it).

With different timeframes—depending on the complexity of the work and the speed of Administrations—former slaughterhouses, ancient monasteries, abandoned industrial sheds, disused schools, historical buildings and other public properties were recovered to become spaces for art, entertainment, the recovery of traditions and ancient crafts; places of social use and experimentation of new technologies; services for work, education and youth entrepreneurship; exhibition, socialisation and hospitality spaces.

With more than 100,000 square meters of renovated area and a total of 54 million Euros—44 million of which came from the 'Framework Programme Agreement on Youth Policy' and 10 million originating from co-financing by the beneficiary municipalities—the Laboratori Urbani action is the largest infrastructure initiative for young people ever carried out by an Italian region. In the Urban Laboratories, the activities are carried out by associations or companies responsible for managing the spaces, selected by the administrations that own the properties through public procedures (Fig. 9.1).

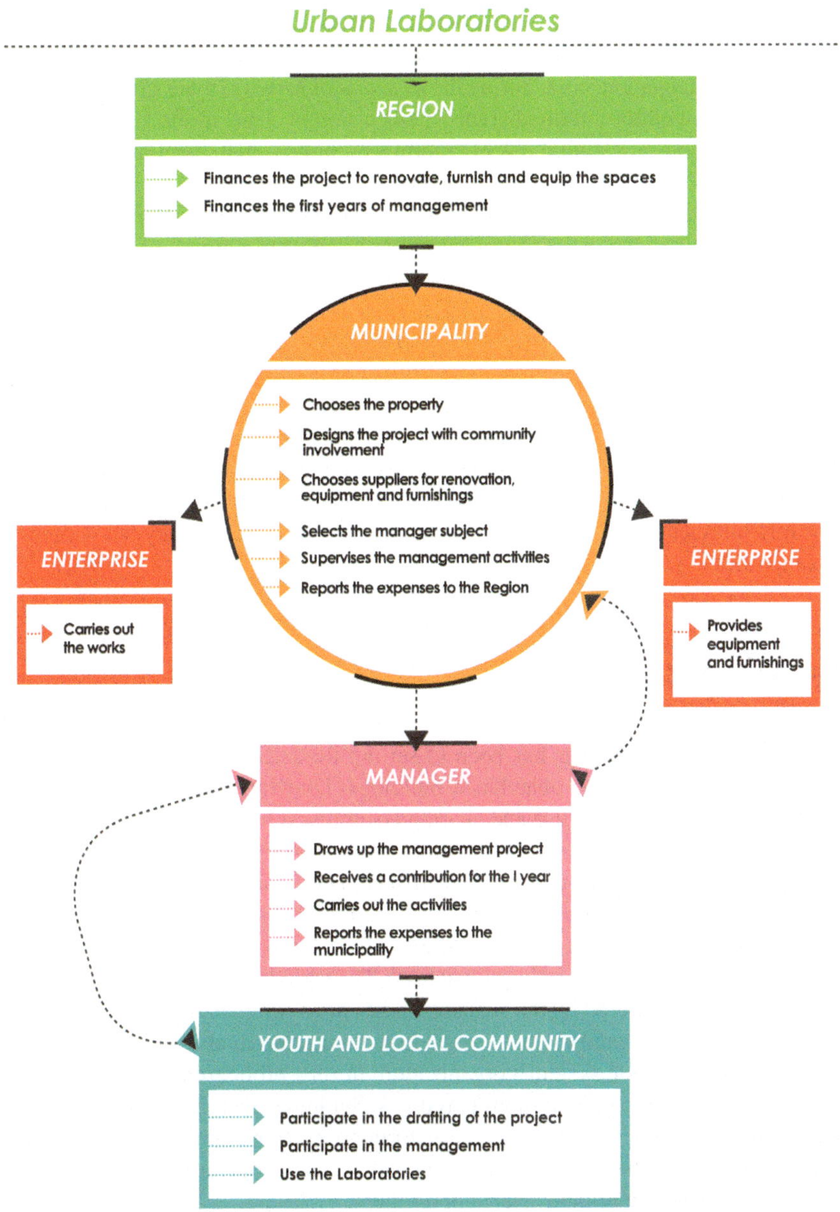

Fig. 9.1 Organizational structure of "Laboratori Urbani" (Urban Laboratories). *Source* Own elaboration from metticilemani.it

A very important factor in these policies is the decision to provide small loans (up to a maximum of 700,000 Euros per project) to restructure and equip the spaces and to incur the management costs during the first year of activity: loans given directly to entrepreneurs, without intermediaries. The decision is clear: in the knowledge that these interventions are at high risk of "mortality", only a small amount of funding is provided, but for some young entrepreneurs, it will be sufficient.

In this way, it really is possible to promote processes of opening up spaces to the vital forces of the territory, experimenting with innovative forms of management, strengthening infrastructure and instrumentation and diversifying activities through the involvement of grassroots associations.

To conclude, we must reflect briefly on the rules. The law can be considered as an infrastructure between "hard" and "soft": the former, as it allows something to happen, and in doing so it creates or destroys value; the latter, as it is not physical. What has emerged from all the cases, from the meeting with all the developers and with the Public Administrations, is the need for simple basic things: the idea of smart actions being something complex should be forgotten. In order to establish a creative city, it is sufficient to allow people be creative without a target policy but with more flexible rules.

9.5 Conclusions

The purpose of this chapter has been to set out the background to conditions for activating the reuse of the abandoned buildings. Drawing on the concepts of circular economy, adaptive reuse and sharing economy, it has been discusses first the opportunity (or not) to start reuse operations and enhancement of unused real estate; second the types of values that can be activated, and finally the operational aspects that involve the architectural features of the buildings ("shape"), the framework of rules ("rules"), and the soft and hard aspects of the assets that must be managed for giving rise to successful operations ("values"). The varied experiences of reactivating abandoned spaces offered by the six case studies have stimulated general reflections for policies to be adopted in the short and medium term, in absence of a profound reform of the economic system's organization. In zero-profit areas, a new attitude is needed with respect to interventions to reuse the building stock. The attention should shift from the final outcome of the restructuring operations (quantity, use, morphological aspects), to the conditions that can activate these operations: not conditions that strangle them but policies that support them.

References

Abastante, F., & Lami, I. M. (2018). An integrated assessment framework for the requalification of districts facing urban and social decline. In G. Mondini, E. Fattinnanzi, A. Oppio, M. Bottero, & S. Stanghellini (Eds.), *Integrated evaluation for the management of contemporary cities. Green energy and technology*. Cham: Springer.

Abastante, F., Lami, I. M., Mecca, B. (2020). How to revitalise a historic district: A stakeholders-oriented assessment framework of adaptive reuse. In G. Mondini, A. Oppio, S. Stanghellini, M. Bottero, & F. Abastante (Eds.), *Values and functions for future cities. Green energy and technology*. Cham: Springer.

Botsman, R., & Rogers, R. (2010). *What's mine is yours. The rise of collaborative consumption*. New York: Harper Collins Publishers.

Bullen, P. A., & Love, P. E. D. (2011a). Factors influencing the adaptive re-use of buildings. *Journal of Engineering, Design and Technology, 9*(1), 32–46.

Bullen, P. A., & Love, P. E. D. (2011b). Adaptive reuse of heritage buildings. *Structural Survey, 29*(5), 411–421.

Conejos, S., Yung, E. H. K., & Chan, E. H. W. (2014). Evaluation of urban sustainability and adaptive reuse of built heritage areas: A case study on conservation in Hong Kong's CBD. *Journal of Design Research, 12*(4), 260–279.

Davidson, N. M., & Infranca, J. J. (2016). The sharing economy as an urban phenomenon. *Yale Law & Policy Review, 34,* 215.

Dewiyana, E., Ibrahim, N., & Hidayah, H. N. (2016). The green aspects of adaptive reuse of Hotel Penaga. *Procedia—Social and Behavioural Sciences, 222,* 631–643.

Douglas, J. (2006). *Building adaptation* (2nd ed.). Oxford: Butterworth-Heinemann.

Ellen MacArthur Foundation, McKinsey Centre for Business and Environment. (2013). *Towards the circular economy*. Report published by Ellen MacArthur Foundation. https://www.ellenmacarthurfoundation.org/assets/downloads/publications/Ellen-MacArthur-Foundation-Towards-the-Circular-Economy-vol.1.pdf. Accessed July 25, 2019.

Ellen MacArthur Foundation, McKinsey Centre for Business and Environment. (2017). *Cities in the circular economy: An initial exploration*. Report published by Ellen MacArthur Foundation. https://www.ellenmacarthurfoundation.org/assets/downloads/publications/Cities-in-the-CE_An-Initial-Exploration.pdf. Accessed July 25, 2019.

Ellen MacArthur Foundation, SUN, McKinsey Centre for Business and Environment. (2015). *Growth within: A circular economy vision for a competitive Europe*. Report published by Ellen MacArthur Foundation. https://www.ellenmacarthurfoundation.org/assets/downloads/publications/EllenMacArthurFoundation_Growth-Within_July15.pdf. Accessed July 25, 2019.

European Commission. (2019, Summer). *Economic forecast—Italy*. https://ec.europa.eu/info/business-economy-euro/economic-performance-and-forecasts/economic-performance-country/italy/economic-forecast-italy_en. Accessed June 4, 2019.

Ferilli, G., Sacco, P. L., Tavano Blessi, G. (2011). Culture, cities and creative hubs: From the instrumental to the functional value of culture in contemporary local development. In L. Fusco Girard & P. Nijkamp (Ed.), *Sustainable and creative cities*. Farnham: Ashgate.

Fiorani, D., Loughlin, K., & Musso, S. F. (2017). Conservation-adaptation. Keeping alive the spirit of the place. Adaptive reuse of heritage with symbolic value. In *EAAE Transactions on Architectural Education* (Vol. 65), Hasselt, Belgium.

Günçea, K., & Mısırlısoya, D. (2015). Questioning the adaptive reuse of Industrial Heritage and Its Interventions in the Context of Sustainability. *Sociology, 5*(9), 718–727.

ISTAT. (2019). Rapporto Annuale 2019, La Situazione Del Paese. https://www.istat.it/storage/rapporto-annuale/2019/Sintesi2019.pdf. Accessed June 4, 2019.

Landry, C. (2008). *The creative city: A toolkit for urban innovators*. London: Earthscan Publications.

Langston, C. A. (2010). Green adaptive reuse: Issues and strategies for the built environment. In D. Wu, M. Xu, Z. Zhang, & Y. Fang (Eds.), *Proceedings of the 1st International Conference on Sustainable Construction & Risk Management (ICSCRM 2010)* (pp. 1165–1173). China: Chongqing Jiaotong University.

Kirchherr, J., Reike, D., & Hekkert, M. (2017). Conceptualizing the circular economy: An analysis of 114 definitions. *Resources, Conservation and Recycling, 127,* 221–232.

Kohler, N., & Yang, W. (2007). Long-term management of building stocks. *Building Research and Information, 35*(4), 351–362.

McKinsey Global Institute Report. (2016). Available at www.mckinsey.com/featured-insights/urbanization/urban-world-meeting-the-demographic-challenge-in-cities. Accessed May 15, 2019.

Németh, J., & Langhorst, J. (2013). Rethinking urban transformation: Temporary uses for vacant land. *Cities, 40,* 143–150.

Orefice, M. (2007). *Estimo civile* (Voll. 1 e 2). Torino: Utet.

Plevoets, B., & Van Cleempoel, K. (2011). Adaptive reuse as a strategy towards conservation of cultural heritage: A literature review. *Structural Studies, Repairs and Maintenance of Heritage Architecture XII, 118,* 155–163.

Plevoets, B., & Van Cleempoel, K. (2013). Adaptive reuse as an emerging discipline: An historic survey. In G. Cairns (Ed.), *Reinventing architecture and interiors: A socio-political view on building adaptation* (pp. 13–32). London: Libri Publishers.

Remøy, H. T., & Van der Voordt, D. J. M. (2007). A new life: Conversion of vacant office buildings into housing. *Facilities, 25*(3–4), 88–103.

Robiglio, M. (2016). *The adaptive reuse toolkit. How cities can turn their industrial legacy into infrastructure for innovation and growth.* Urban and Regional Policy Paper, 38.

Robiglio, M. (2017). *RE–USA: 20 American stories of adaptive reuse: A toolkit for post-industrial cities.* Berlin: Jovis Verga GmbH

Sacco, P. L. (2017). The essence of smart specialization: Local economies as computational platforms. *Regional Studies, 51*(5), 814–816.

Sacco, P. L., & Tavano Blessi, G. (2009). The social viability of culture-led urban transformation processes: Evidence from the Bicocca District. *Milan Urban Studies, 46,* 1115.

Swart, J., & Kinnie, N. (2003). Sharing knowledge in knowledge-intensive. *Human Resource Management Journal, 13*(2), 60–75.

Throsby, D. (1995). Culture, economics and sustainability. *Journal of Cultural Economics, 19,* 199–206.

Throsby, D. (1999). Cultural capital. *Journal of Cultural Economics, 23,* 3–12.

Williams, J. (2019). Circular cities, critical commentaries. *Urban Studies.*

www.metticilemani.it. Laboratori urbani metticilemani. Accessed June 3, 2019.